SpringerBriefs in Mathematics

Series Editors

Krishnaswami Alladi
Nicola Bellomo
Michele Benzi
Tatsien Li
Matthias Neufang
Otmar Scherzer
Dierk Schleicher
Vladas Sidoravicius
Benjamin Steinberg
Yuri Tschinkel
Loring W. Tu
G. George Yin
Ping Zhang

T0178348

SpringerBriefs in Mathematics showcases expositions in all areas of mathematics and applied mathematics. Manuscripts presenting new results or a single new result in a classical field, new field, or an emerging topic, applications, or bridges between new results and already published works, are encouraged. The series is intended for mathematicians and applied mathematicians.

For further volumes:
http://www.springer.com/series/10030

Kurt Luoto • Stefan Mykytiuk
Stephanie van Willigenburg

An Introduction to Quasisymmetric Schur Functions

Hopf Algebras, Quasisymmetric Functions, and Young Composition Tableaux

Springer

Kurt Luoto
Department of Mathematics
University of British Columbia
Vancouver, British Columbia
Canada

Stefan Mykytiuk
Department of Mathematics and Statistics
York University
Toronto, Ontario
Canada

Stephanie van Willigenburg
Department of Mathematics
University of British Columbia
Vancouver, British Columbia
Canada

ISSN 2191-8198 ISSN 2191-8201 (electronic)
ISBN 978-1-4614-7299-5 ISBN 978-1-4614-7300-8 (eBook)
DOI 10.1007/978-1-4614-7300-8
Springer New York Heidelberg Dordrecht London

Library of Congress Control Number: 2013936258

Mathematics Subject Classification (2010): 05A05, 05E05, 05E15, 06A07, 16T05, 16T30

© Kurt Luoto, Stefan Mykytiuk, Stephanie van Willigenburg 2013
This work is subject to copyright. All rights are reserved by the Publisher, whether the whole or part of the material is concerned, specifically the rights of translation, reprinting, reuse of illustrations, recitation, broadcasting, reproduction on microfilms or in any other physical way, and transmission or information storage and retrieval, electronic adaptation, computer software, or by similar or dissimilar methodology now known or hereafter developed. Exempted from this legal reservation are brief excerpts in connection with reviews or scholarly analysis or material supplied specifically for the purpose of being entered and executed on a computer system, for exclusive use by the purchaser of the work. Duplication of this publication or parts thereof is permitted only under the provisions of the Copyright Law of the Publisher's location, in its current version, and permission for use must always be obtained from Springer. Permissions for use may be obtained through RightsLink at the Copyright Clearance Center. Violations are liable to prosecution under the respective Copyright Law.
The use of general descriptive names, registered names, trademarks, service marks, etc. in this publication does not imply, even in the absence of a specific statement, that such names are exempt from the relevant protective laws and regulations and therefore free for general use.
While the advice and information in this book are believed to be true and accurate at the date of publication, neither the authors nor the editors nor the publisher can accept any legal responsibility for any errors or omissions that may be made. The publisher makes no warranty, express or implied, with respect to the material contained herein.

Printed on acid-free paper

Springer is part of Springer Science+Business Media (www.springer.com)

For Niall Christie and Madge Luoto

Preface

The history of quasisymmetric functions begins in 1972 with the thesis of Richard Stanley, followed by the formal definition of the Hopf algebra of quasisymmetric functions in 1984 by Ira Gessel. From this definition a whole research area grew and a more detailed, although not exhaustive, history can be found in the introduction.

The history of quasisymmetric Schur functions is far more contemporary. They were discovered in 2007 during the semester on "Recent Advances in Combinatorics" at the Centre de Recherches Mathématiques, and further progress was made at a variety of workshops at the Banff International Research Station and during an Alexander von Humboldt Foundation Fellowship awarded to Steph. The idea for writing this book came from encouragement by Adriano Garsia who suggested we recast quasisymmetric Schur functions using tableaux analogous to Young tableaux. We followed his words of wisdom.

The aim of this monograph is twofold. The first goal is to provide a reference text for the basic theory of Hopf algebras, in particular the Hopf algebras of symmetric, quasisymmetric and noncommutative symmetric functions and connections between them. The second goal is to give a survey of results with respect to an exciting new basis of the Hopf algebra of quasisymmetric functions, whose combinatorics is analogous to that of the renowned Schur functions.

In particular, after introducing the topic in Chapter 1, in Chapter 2 we review pertinent combinatorial concepts such as partially ordered sets, Young and reverse tableaux, and Schensted insertion. In Chapter 3 we give the basic theory of Hopf algebras, illustrating it with the Hopf algebras of symmetric, quasisymmetric and noncommutative symmetric functions, ending with a brief introduction to combinatorial Hopf algebras. The exposition is based on Stefan's thesis, useful personal notes made by Kurt, and a talk Steph gave entitled "Everything you wanted to know about Sym, QSym and NSym but were afraid to ask". Chapter 4 generalizes concepts from Chapter 2 such as Young tableaux and reverse tableaux indexed by partitions, to Young composition tableaux and reverse composition tableaux indexed by compositions. The final chapter then introduces two natural refinements for the Schur functions from Chapter 3: quasisymmetric Schur functions reliant on reverse composition tableaux and Young quasisymmetric Schur functions reliant on Young composition tableaux. This chapter concludes by discussing a number of results for these Schur function refinements and their dual bases. These results are analogous to those found in the theory of Schur functions such as the computation of Kostka numbers, and Pieri and Littlewood–Richardson rules. Throughout parallel construction is used so that analogies may easily be spotted even when browsing.

None of this would be possible without the support of a number of people, whom we would now like to thank. Firstly, Adriano Garsia has our sincere thanks for his ardent support of pursuing quasisymmetric Schur functions. We are also grateful to our advisors and mentors who introduced us to, and fuelled our enthusiasm for, quasisymmetric functions: Nantel Bergeron, Lou Billera, Sara Billey, Isabella Novik and Frank Sottile. This enthusiasm was sustained by our coauthors on our papers involving quasisymmetric Schur functions: Christine Bessenrodt, Jim Haglund and Sarah Mason, with whom it was such a pleasure to do research. We are also fortunate to have visited a variety of stimulating institutes to conduct our research and our thanks go to Francois Bergeron and a host of enthusiastic colleagues who arranged the aforementioned semester. Plus we are most grateful for the opportunities at Banff afforded to us by the director of BIRS, Nassif Ghoussoub, his team, the organizers of each of the meetings we attended and the participants all of whom gave us a stimulating and supportive atmosphere for us to pursue our goals. We are also grateful to the reviewers of this book, to Ole Warnaar, and to Moss Sweedler for their advice.

Our various universities, York University, the University of Washington and the University of British Columbia, are thanked for their support, in particular the A+E Combinatorics reading group at the latter: Omer Angel, Caleb Cheek, Andrew Rechnitzer, Tom Wong, and especially Ed Richmond and Vasu Tewari, both of whom kindly agreed to proofread this manuscript. No research is possible without funding, and we are grateful to be supported by the Natural Sciences and Engineering Research Council of Canada and the Alexander von Humboldt Foundation.

We would also like to thank Razia Amzad at Springer US for her help and support during the preparation of this manuscript, and our families and friends for all their love and support. Lastly, we would like to thank you, the reader, who we hope finds our book a rewarding read.

Vancouver, BC, Canada — Kurt Luoto
Toronto, ON, Canada — Stefan Mykytiuk
Vancouver, BC, Canada — Stephanie van Willigenburg

Notation

α	Composition
$\tilde{\alpha}$	Underlying partition of α
α^c	Complement of α
α^r	Reversal of α
α^t	Transpose of α
$\alpha /\!\!/_{\check{c}} \beta$	Skew shape
$\alpha /\!\!/_{\hat{c}} \beta$	Skew shape
col	Column sequence of a tableau
comp	Composition corresponding to a subset, or to a descent set of a tableau
cont	Content of a tableau
χ	Forgetful map
d	Descent set of a permutation
D	Descent set of a chain
Des	Descent set of a tableau
δ_{ij}	1 if $i = j$ and 0 otherwise
Δ	Coalgebra coproduct
e_λ	Elementary symmetric function
$\mathbf{e}_\alpha$	Elementary noncommutative symmetric function
F_α	Fundamental quasisymmetric function
$\mathscr{H}$	Hopf algebra
$\mathscr{H}^*$	Dual Hopf algebra
h_λ	Complete homogeneous symmetric function
$\mathbf{h}_\alpha$	Complete homogeneous noncommutative symmetric function
ℓ	Length of a composition or partition
$\mathscr{L}$	Set of linear extensions of a poset

$\mathcal{L}_{\check{c}}$	Reverse composition poset
$\mathcal{L}_{\hat{c}}$	Young composition poset
$\mathcal{L}_Y$	Young's lattice
λ	Partition
λ^t	Transpose of λ
λ/μ	Skew shape
m_λ	Monomial symmetric function
M_α	Monomial quasisymmetric function
$[n]$	The set $\{1,2,\ldots,n\}$
NSym	Hopf algebra of noncommutative symmetric functions
P	Poset
P^*	Dual poset
P	P-tableau of a list
QSym	Hopf algebra of quasisymmetric functions
r_α	Ribbon Schur function
$\mathbf{r}_\alpha$	Noncommutative ribbon Schur function
$\check{\rho}_\alpha$	Bijection between SSRCT and SSRT
$\hat{\rho}_\alpha$	Bijection between SSYCT and SSYT
S	Antipode
set	Set corresponding to a composition
sh	Shape of a tableau
s_λ	Schur function
$\check{\mathcal{S}}_\alpha$	Quasisymmetric Schur function
$\hat{\mathcal{S}}_\alpha$	Young quasisymmetric Schur function
$\check{\mathbf{s}}_\alpha$	Noncommutative Schur function
$\hat{\mathbf{s}}_\alpha$	Young noncommutative Schur function
SRT	Standard reverse tableau
$SRCT$	Standard reverse composition tableau
$SSRT$	Semistandard reverse tableau
$SSRCT$	Semistandard reverse composition tableau
Sym	Hopf algebra of symmetric functions
SYT	Standard Young tableau
$SYCT$	Standard Young composition tableau
$SSYT$	Semistandard Young tableau
$SSYCT$	Semistandard Young composition tableau
$\mathfrak{S}_n$	Symmetric group
$\check{T}$	Reverse tableau
T	Young tableau

$\check{\tau}$	Reverse composition tableau
τ	Young composition tableau
$\check{U}_\alpha$	Canonical standard reverse composition tableau
U_α	Canonical standard Young composition tableau
$\check{V}_\lambda$	Canonical standard reverse tableau
V_λ	Canonical standard Young tableau
w_{col}	Column reading word of a tableau
x^T	Monomial of a tableau
$\emptyset$	Empty composition
$\vdash$	Is a partition of
$\vDash$	Is a composition of
$\lvert\ \rvert$	Weight of a composition or partition, or size of a skew shape
$\cdot$	Concatenation of compositions
$\odot$	Near concatenation of compositions
$\sqcup\!\sqcup$	Shuffle of permutations
$+$	Disjoint union of posets
$\lessdot$	Cover relation in a poset
$\leqslant$	Relation in a poset
$\preccurlyeq$	Refines
$\subseteq$	Subset of, or containment
$[\ ,\ ]$	Closed interval in a poset
$(\ ,\ )$	Open interval in a poset
$\langle\ ,\ \rangle$	Bilinear form, or inner product

Contents

Chapter 1
Introduction

Abstract A brief history of the Hopf algebra of quasisymmetric functions is given, along with their appearance in discrete geometry, representation theory and algebra. A discussion on how quasisymmetric functions simplify other algebraic functions is undertaken, and their appearance in areas such as probability, topology, and graph theory is also covered. Research on the dual algebra of noncommutative symmetric functions is touched on, as is a variety of extensions to quasisymmetric functions. What is known about the basis of quasisymmetric Schur functions is also addressed.

1.1 A brief history of quasisymmetric functions

We begin with a brief history of quasisymmetric functions ending with the recent discovery of quasisymmetric Schur functions, which will give an indication of the depth and breadth of this fascinating subject.

The history starts with plane partitions that were discovered by MacMahon [61] and later connected to the theory of symmetric functions by, for example, Bender and Knuth [9]. MacMahon's work anticipated the theory of P-partitions, which was first developed explicitly by Stanley [77] in 1972, and laid out the basic theory of quasisymmetric functions in this context, but not in the language of quasisymmetric functions. The definition of quasisymmetric functions was given in 1984 by Gessel [35] who also described many of the fundamental properties of the Hopf algebra of quasisymmetric functions, Qsym. Ehrenborg [27] developed further Hopf algebraic properties of quasisymmetric functions, and employed them to encode the flag f-vector of a poset, meanwhile proving that Qsym is dual to Solomon's descent algebra of type A was done by Malvenuto and Reutenauer [62] in 1995. It was at this point in time that the study of Qsym and related algebras began to fully blossom.

The theory of descent algebras of Coxeter groups [76] was already a rich subject in type A, for example [32], and in [34] Solomon's descent algebra of type A was shown to be isomorphic to the Hopf algebra of noncommutative symmetric functions Nsym. This latter algebra is isomorphic [41] to the universal enveloping

K. Luoto et al., *An Introduction to Quasisymmetric Schur Functions*,
SpringerBriefs in Mathematics, DOI 10.1007/978-1-4614-7300-8_1,

© Kurt Luoto, Stefan Mykytiuk, Stephanie van Willigenburg 2013

algebra of the free Lie algebra with countably many generators [71] and formed the basis of another fruitful avenue of research, for example, [26, 34, 51]. This avenue led to extending noncommutative symmetric functions to more than one parameter [54], coloured trees [85] and noncommutative character theory [21].

With strong connections to discrete geometry [27,35] quasisymmetric functions also arise frequently in areas within discrete geometry such as in the study of posets [53,66,80,83], combinatorial polytopes [22], matroids [19,24,59] and the **cd**-index [17]. Plus there is a natural strong connection to algebra, and the Hopf algebra of quasisymmetric functions is isomorphic to the Hopf algebra of ladders [30], is free over the Hopf algebra of symmetric functions [33], and can have its polynomial generators computed [44]. Further algebraic properties can be found in [42, 43]. Qsym is also the terminal object in the category of combinatorial Hopf algebras [4], and facilitates the computation of their characters [67]. Meanwhile, in the context of representation theory, quasisymmetric functions arise in the study of Hecke algebras [46], Lie representations [36], crystal graphs for general linear Lie superalgebras [52], and explicit formulas for the odd and even parts of the universal character on Qsym are given in [2]. However, it is arguable that quasisymmetric functions have had the greatest impact in simplifying the computation of many well-known functions. Examples include Macdonald polynomials [38,40], skew Hall-Littlewood polynomials [58], Kazhdan-Lusztig polynomials [16], Stanley symmetric functions [78], shifted quasisymmetric functions [13,20] and the plethysm of Schur functions [57]. Many other examples arise through the theory of Pieri operators on posets [12].

Quasisymmetric functions also arise in the study of Tchebyshev transforms where the Tchebyshev transform of the second kind is a Hopf algebra endomorphism on Qsym [28] used as a tool in establishing Schur positivity [5]. With respect to ribbon Schur functions, the sum of fundamental quasisymmetric functions over a forgotten class is a multiplicity free sum of ribbon Schur functions [70], while fundamental quasisymmetric functions are key to determining when two ribbon Schur functions are equal [18]. In graph theory, quasisymmetric functions can be used to describe the chromatic symmetric function [23, 79], and recently quasisymmetric refinements of the chromatic symmetric function have been introduced [50, 75]. In enumerative combinatorics, quasisymmetric functions are combined with the statistics of major index and excedance to create Eulerian quasisymmetric functions [74] although these functions are, in fact, symmetric. The topology of Qsym has been studied in [7, 37] and its impact on probability comes via the study of riffle shuffles [82] and random walks [45]. Quasisymmetric functions also play a role in the study of trees [47, 87], and the KP hierarchy [25].

Generalizations and extensions of Qsym are also numerous and include the Malvenuto-Reutenauer Hopf algebra of permutations, denoted by $\mathfrak{S}$ Sym or FQSym, for example [3, 26, 62]; quasisymmetric functions in noncommuting variables, for example [10]; higher level quasisymmetric functions, for example [48]; coloured quasisymmetric functions, for example [49]; Type B quasisymmetric functions, for example [8]; and the space R_n constructed as a quotient by the ideal of quasisymmetric polynomials with no constant term, for example [6].

Recently, a new basis of Qsym has been discovered: the basis of quasisymmetric Schur functions [40], which arises from the combinatorics of Macdonald polynomials [38]. Schur functions are a basis for the Hopf algebra of symmetric functions, Sym, and are a central object of study due to their omnipresent nature: from being generating functions for tableaux to being irreducible characters of the general linear groups. Their ubiquity is well documented in classic texts such as [31, 60, 72, 81]. Quasisymmetric Schur functions refine Schur functions in a natural way and, moreover, exhibit many of the elegant properties of Schur functions. Properties include exhibiting quasisymmetric Kostka numbers [40] and Littlewood-Richardson rules [15, 39], while their image under the involution ω yields row-strict quasisymmetric Schur functions [29, 65]. Additionally quasisymmetric Schur functions have had certain multiplicity free expansions computed [14], and were pivotal in resolving a conjecture of F. Bergeron and Reutenauer that Qsym over Sym has a stable basis [55]. They can also be used to prove Schur positivity in two stages: if a function is proved to be both quasisymmetric Schur positive and symmetric, then it is Schur positive.

While much has been done, there is still, without doubt, a plethora of theorems to discover about quasisymmetric Schur functions.

Chapter 2
Classical combinatorial concepts

Abstract In this chapter we begin by defining partially ordered sets, linear extensions, the dual of a poset, and the disjoint union of two posets. We then define further combinatorial objects we will need including compositions, partitions, diagrams and Young tableaux, reverse tableaux, Young's lattice and Schensted insertion.

2.1 Partially ordered sets

A useful notion for us throughout this book will be that of a partially ordered set.

Definition 2.1.1. A *partially ordered set*, or simply *poset*, is a pair $(P, \leqslant)$ consisting of a set P and a binary relation $\leqslant$ on P that is reflexive, antisymmetric and transitive, that is, for all $p, q, r \in P$,

1. $p \leqslant p$
2. $p \leqslant q$ and $q \leqslant p$ implies $p = q$
3. $p \leqslant q$ and $q \leqslant r$ implies $p \leqslant r$.

The relation $\leqslant$ is called a *partial order* on or *partial ordering* of P.

We write $p < q$ if $p \leqslant q$ and $p \neq q$, $p \geqslant q$ if $q \leqslant p$, and $p > q$ if $q < p$. Elements $p, q \in P$ are called *comparable* if $p \leqslant q$ or $q \leqslant p$.

If $p \leqslant q$, then we define the *closed interval*

$$[p,q] = \{r \in P | p \leqslant r \leqslant q\}$$

and the *open interval*

$$(p,q) = \{r \in P | p < r < q\}.$$

An element q *covers* an element p if $p < q$ and $(p,q) = \emptyset$. If q covers p, then we write $p \lessdot q$.

K. Luoto et al., *An Introduction to Quasisymmetric Schur Functions*, SpringerBriefs in Mathematics, DOI 10.1007/978-1-4614-7300-8_2,
© Kurt Luoto, Stefan Mykytiuk, Stephanie van Willigenburg 2013

A *chain* is a poset in which any two elements are comparable. The order here is called a *total* or *linear* order. A *saturated chain* of *length* $n-1$ in a poset is a subset with order $q_1 \lessdot \cdots \lessdot q_n$. For our purposes we say a poset is *graded* if it has a unique minimal element $\hat{0}$ and every saturated chain between $\hat{0}$ and a poset element x has the same length, called the *rank* of x.

Example 2.1.2. Familiar posets include $(\mathbb{Z}, \leqslant)$ where $\leqslant$ is the usual relation *less than or equal to* on the integers, and $(\mathscr{P}(A), \subseteq)$ where $\mathscr{P}(A)$ is the collection of all subsets of a set A. In addition, the poset $(\mathbb{Z}, \leqslant)$ is a chain.

We shall often abuse notation and give both a poset and its underlying set the same name. Thus a *poset* P shall mean, unless otherwise specified, a set P together with a partial order on P. The partial order will usually be denoted by the symbol $\leqslant$, with the words 'in P', or subscript P, added if necessary to distinguish it from the partial order on a different poset.

We will be interested in chains that contain a given poset in the following sense.

Definition 2.1.3. A *linear extension* of a poset P is a chain w consisting of the set P with a total order that satisfies

$$p < q \text{ in } P \text{ implies } p < q \text{ in } w.$$

When P is finite, we can let this total order be $w_1 < w_2 < \cdots$ and restate the last condition as

$$w_i < w_j \text{ in } P \text{ implies } i < j.$$

The set of all linear extensions of P is denoted by $\mathscr{L}(P)$.

Example 2.1.4. There are two linear extensions of $(\mathscr{P}(\{1,2\}), \subseteq)$, namely

$$\emptyset < \{1\} < \{2\} < \{1,2\}$$

and

$$\emptyset < \{2\} < \{1\} < \{1,2\}.$$

Finally, we introduce two operations on posets. The *dual* of a poset P is the poset P^* consisting of the set P with partial order defined by

$$p \leqslant q \text{ in } P^* \text{ if } q \leqslant p \text{ in } P.$$

If P and Q are posets with disjoint underlying sets P and Q, then the *disjoint union* $P+Q$ is the poset consisting of the set $P \cup Q$ with partial order defined by

$$p \leqslant q \text{ in } P+Q \text{ if } p \leqslant q \text{ in } P \text{ or } p \leqslant q \text{ in } Q.$$

Since P and Q are disjoint, $p \leqslant q$ in $P+Q$ is possible only if $p,q \in P$ or $p,q \in Q$.

2.2 Compositions and partitions

Compositions and partitions will be the foundation for the indexing sets of the functions we will be studying.

Definition 2.2.1. A *composition* is a finite ordered list of positive integers. A *partition* is a finite unordered list of positive integers that we write in weakly decreasing order when read from left to right. In both cases we call the integers the *parts* of the composition or partition.

The *underlying partition* of a composition α, denoted by $\tilde{\alpha}$, is the partition obtained by sorting the parts of α into weakly decreasing order.

Given a composition or partition $\alpha = (\alpha_1, \ldots, \alpha_k)$, we define its *weight* or *size* to be $|\alpha| = \alpha_1 + \cdots + \alpha_k$ and its *length* to be $\ell(\alpha) = k$. When $\alpha_{j+1} = \cdots = \alpha_{j+m} = i$ we often abbreviate this sublist to i^m. If α is a composition with $|\alpha| = n$, then we write $\alpha \vDash n$ and say α is a composition of n. If λ is a partition with $|\lambda| = n$, then we write $\lambda \vdash n$ and say λ is a partition of n. For convenience we denote by $\emptyset$ the unique composition or partition of weight and length 0, called the *empty* composition or partition.

Example 2.2.2. The compositions of 4 are

$$(4), (3,1), (1,3), (2,2), (2,1,1), (1,2,1), (1,1,2), (1,1,1,1).$$

The partitions of 4 are

$$(4), (3,1), (2,2), (2,1,1), (1,1,1,1).$$

If $\alpha = (1,4,1,2)$, then $\tilde{\alpha} = (4,2,1,1)$.

Let $[n] = \{1, 2, \ldots, n\}$. There is a natural one-to-one correspondence between compositions of n and subsets of $[n-1]$, given by the following.

Definition 2.2.3. Let n be a nonnegative integer.

1. If $\alpha = (\alpha_1, \ldots, \alpha_k) \vDash n$, then we define

$$\mathrm{set}(\alpha) = \{\alpha_1, \alpha_1 + \alpha_2, \ldots, \alpha_1 + \cdots + \alpha_{k-1}\} \subseteq [n-1].$$

2. If $A = \{a_1, \ldots, a_\ell\} \subseteq [n-1]$ where $a_1 < \cdots < a_\ell$, then we define

$$\mathrm{comp}(A) = (a_1, a_2 - a_1, \ldots, a_\ell - a_{\ell-1}, n - a_\ell) \vDash n.$$

In particular, the empty set corresponds to the composition $\emptyset$ if $n = 0$, and to (n) if $n > 0$.

Example 2.2.4. Let $\alpha = (1,1,3,1,2) \vDash 8$. Then

$$\text{set}(\alpha) = \{1, 1+1, 1+1+3, 1+1+3+1\} = \{1,2,5,6\} \subseteq [7].$$

Conversely, if $A = \{1,2,5,6\} \subseteq [7]$, then

$$\text{comp}(A) = (1, 2-1, 5-2, 6-5, 8-6) = (1,1,3,1,2).$$

For every composition $\alpha \vDash n$ there exist three closely related compositions: its reversal, its complement, and its transpose. Firstly, the *reversal* of α, denoted by α^r, is obtained by writing the parts of α in the reverse order. Secondly, the *complement* of α, denoted by α^c, is given by

$$\alpha^c = \text{comp}(\text{set}(\alpha)^c),$$

that is, α^c is the composition that corresponds to the complement of the set that corresponds to α. Lastly, the *transpose* (also known as the *conjugate*) of α, denoted by α^t, is defined to be $\alpha^t = (\alpha^r)^c = (\alpha^c)^r$.

Example 2.2.5. If $\alpha = (1,4,1,2) \vDash 8$, then $\text{set}(\alpha) = \{1,5,6\} \subseteq [7]$, and hence $\alpha^r = (2,1,4,1)$, $\alpha^c = (2,1,1,3,1)$, $\alpha^t = (1,3,1,1,2)$.

Pictorially, we can view a composition $\alpha = (\alpha_1, \ldots, \alpha_k)$ as a row consisting of α_1 dots, then a bar followed by α_2 dots, then a bar followed by α_3 dots, and so on. We can use the picture of α to create the pictures of $\alpha^r, \alpha^c, \alpha^t$ as follows.

To create the picture of α^r, reflect the picture of α in a vertical axis. To create the picture of α^c, place a bar between two dots if there is no bar between the corresponding dots in the picture of α. Finally, create the picture of α^t by performing one of these actions, then using the resulting picture to perform the other action.

Example 2.2.6. Repeating our previous example, if $\alpha = (1,4,1,2)$ then the picture of α is

$$\bullet \,|\, \bullet\bullet\bullet\bullet \,|\, \bullet \,|\, \bullet\bullet$$

so to compute $\alpha^r, \alpha^c, \alpha^t$ we draw the pictures

$$\bullet\bullet \,|\, \bullet \,|\, \bullet\bullet\bullet\bullet \,|\, \bullet \quad , \quad \bullet\bullet \,|\, \bullet \,|\, \bullet \,|\, \bullet\bullet\bullet \,|\, \bullet \quad , \quad \bullet \,|\, \bullet\bullet\bullet \,|\, \bullet \,|\, \bullet \,|\, \bullet\bullet$$

to obtain $\alpha^r = (2,1,4,1)$, $\alpha^c = (2,1,1,3,1)$, $\alpha^t = (1,3,1,1,2)$.

Given a pair of compositions, there are also two operations that can be performed. The *concatenation* of $\alpha = (\alpha_1, \ldots, \alpha_k)$ and $\beta = (\beta_1, \ldots, \beta_\ell)$ is

$$\alpha \cdot \beta = (\alpha_1, \ldots, \alpha_k, \beta_1, \ldots, \beta_\ell)$$

while the *near concatenation* is

$$\alpha \odot \beta = (\alpha_1, \ldots, \alpha_k + \beta_1, \ldots, \beta_\ell).$$

For example, $(1,4,1,2)\cdot(3,1,1)=(1,4,1,2,3,1,1)$ while $(1,4,1,2)\odot(3,1,1)=(1,4,1,5,1,1)$.

Given compositions α,β, we say that α is a *coarsening* of β (or equivalently β is a *refinement* of α), denoted by $\alpha\succcurlyeq\beta$, if we can obtain the parts of α in order by adding together adjacent parts of β in order. For example, $(1,4,1,2)\succcurlyeq(1,1,3,1,2)$.

We end this section with the following result on refinement, which is straightforward to verify, and is illustrated by Examples 2.2.4 and 2.2.5 and the definition of refinement.

Proposition 2.2.7. *Let α and β be compositions of the same weight. Then*

$$\alpha\preccurlyeq\beta \text{ if and only if } \mathrm{set}(\beta)\subseteq\mathrm{set}(\alpha).$$

2.3 Partition diagrams

We now associate compositions and partitions with diagrams.

Definition 2.3.1. Given a partition $\lambda=(\lambda_1,\ldots,\lambda_{\ell(\lambda)})\vdash n$, we say the *Young diagram* of λ, also denoted by λ, is the left-justified array of n cells with λ_i cells in the i-th row. We follow the Cartesian or French convention, which means that we number the rows from bottom to top, and the columns from left to right. The cell in the i-th row and j-th column is denoted by the pair (i,j).

Example 2.3.2. The cell filled with a $*$ is the cell $(2,3)$.

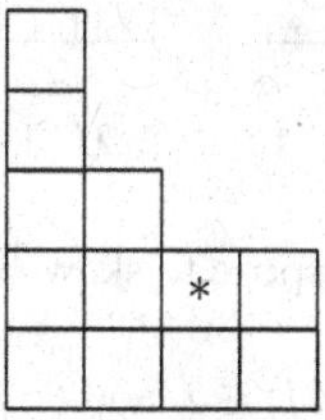

$$\lambda=(4,4,2,1,1)$$

Let λ,μ be two Young diagrams. We say μ is *contained* in λ, denoted by $\mu\subseteq\lambda$, if $\ell(\mu)\leqslant\ell(\lambda)$ and $\mu_i\leqslant\lambda_i$ for $1\leqslant i\leqslant\mu_{\ell(\mu)}$. If $\mu\subseteq\lambda$, then we define the *skew shape* λ/μ to be the array of cells

$$\lambda/\mu=\{(i,j)\mid(i,j)\in\lambda\text{ and }(i,j)\notin\mu\}.$$

For convenience, we refer to μ as the *inner shape* and to λ as the *outer shape*. The *size* of λ/μ is $|\lambda/\mu|=|\lambda|-|\mu|$. Note that the skew shape $\lambda/\emptyset$ is the same as the Young diagram λ. Consequently, we write λ instead of $\lambda/\emptyset$. Such a skew shape is said to be of *straight shape*.

Example 2.3.3. In this example the inner shape is denoted by cells filled with a •, although often these cells are not drawn.

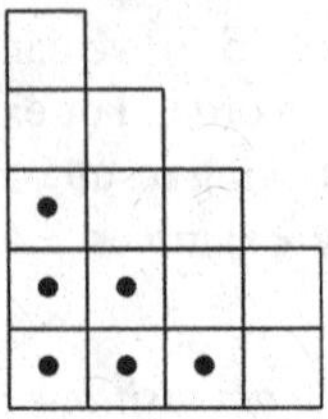

$$\lambda/\mu = (4,4,3,2,1)/(3,2,1)$$

The *transpose* of a Young diagram λ, denoted by λ^t, is the array of cells

$$\lambda^t = \{(j,i) | (i,j) \in \lambda\}.$$

Note that this defines the transpose of a partition.

Example 2.3.4.

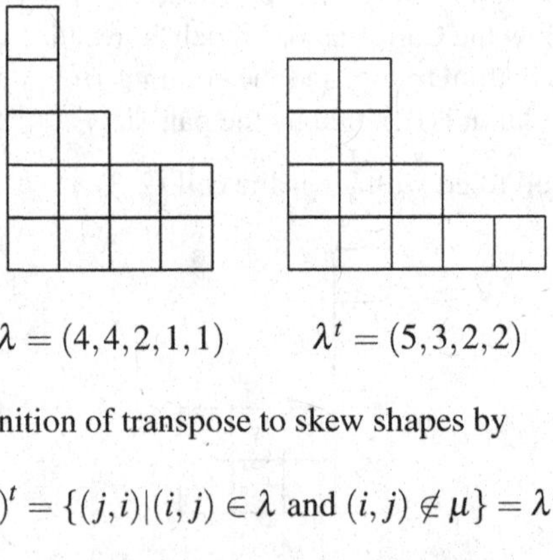

$\lambda = (4,4,2,1,1)$ $\lambda^t = (5,3,2,2)$

We extend the definition of transpose to skew shapes by

$$(\lambda/\mu)^t = \{(j,i) | (i,j) \in \lambda \text{ and } (i,j) \notin \mu\} = \lambda^t/\mu^t.$$

Three skew shapes of particular note are horizontal strips, vertical strips and ribbons. We say a skew shape is a *horizontal strip* if no two cells lie in the same column, and a *vertical strip* if no two cells lie in the same row. A skew shape is *connected* if for every cell d with another cell strictly below or to the right of it, there exists a cell edge-adjacent to d either below or to the right. We say a connected skew shape is a *ribbon* if the following subarray of four cells does not occur in it.

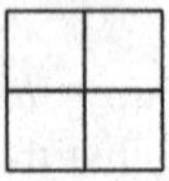

It follows that a ribbon is an array of cells in which, if we number rows from top to bottom, the leftmost cell of row $i+1$ lies immediately below the rightmost cell of row i. Consequently, a ribbon can be efficiently indexed by the composition $\alpha = (\alpha_1, \alpha_2, \ldots, \alpha_{\ell(\alpha)})$, where α_i is the number of cells in row i. This indexing, which follows [34] and involves a deviation from the Cartesian convention for numbering rows, will simplify our discussion of duality later. It also ensures that the definitions of transpose of a ribbon and transpose of a composition agree, as illustrated in Examples 2.2.5 and 2.3.5.

Example 2.3.5.

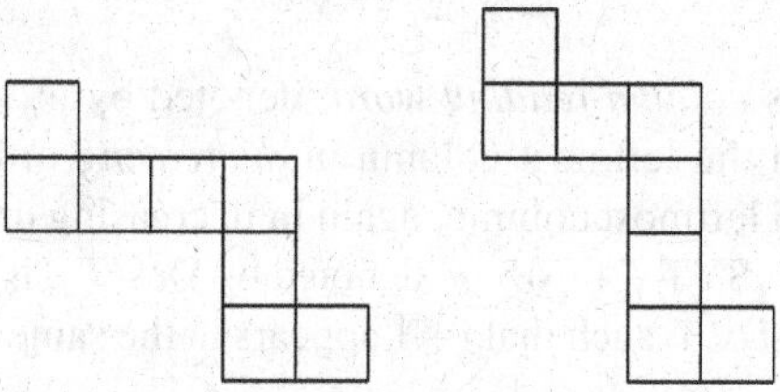

$$\alpha = (1, 4, 1, 2) \qquad \alpha^t = (1, 3, 1, 1, 2)$$

2.4 Young tableaux and Young's lattice

We now take the diagrams of the previous section and fill their cells with positive integers to form tableaux.

Definition 2.4.1. Given a skew shape λ/μ, we define a *semistandard Young tableau* (abbreviated to *SSYT*) T of *shape* $sh(T) = \lambda/\mu$ to be a filling

$$T : \lambda/\mu \to \mathbb{Z}^+$$

of the cells of λ/μ such that

1. the entries in each row are weakly increasing when read from left to right
2. the entries in each column are strictly increasing when read from bottom to top.

A *standard Young tableau* (abbreviated to *SYT*) is an SSYT in which the filling is a bijection $T : \lambda/\mu \to \lceil |\lambda/\mu| \rceil$, that is, each of the numbers $1, 2, \ldots, |\lambda/\mu|$ appears exactly once. Sometimes we will abuse notation and use SSYTs and SYTs to denote the set of all such tableaux.

Example 2.4.2. An SSYT and SYT, respectively, are shown below.

7			
5	5		
•	2	3	
•	•	2	2
•	•	•	1

7			
6	9		
2	5		
1	3	4	8

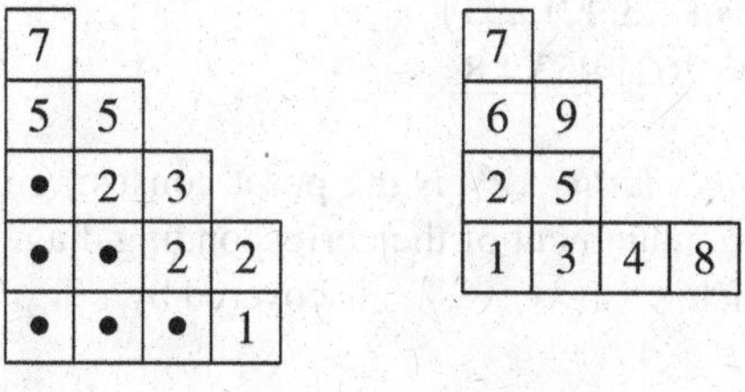

Given an SSYT T, we define the *content* of T, denoted by $\mathrm{cont}(T)$, to be the list of nonnegative integers

$$\mathrm{cont}(T) = (c_1, c_2, \ldots, c_{max})$$

where c_i is the number of times i appears in T, and max is the largest integer appearing in T. Furthermore, given variables $x_1, x_2, \ldots$, we define the *monomial of* T to be

$$x^T = x_1^{c_1} x_2^{c_2} \cdots x_{max}^{c_{max}}.$$

Given an SYT T, its *column reading word*, denoted by $w_{col}(T)$, is obtained by listing the entries from the leftmost column in *decreasing* order, followed by the entries from the second leftmost column, again in decreasing order, and so on.

The *descent set* of an SYT T of size n, denoted by $\mathrm{Des}(T)$, is the subset of $[n-1]$ consisting of all entries i of T such that $i+1$ appears in the same column or a column to the left, that is,

$$\mathrm{Des}(T) = \{i \mid i+1 \text{ appears weakly left of } i\} \subseteq [n-1]$$

and the corresponding *descent composition* of T is

$$\mathrm{comp}(T) = \mathrm{comp}(\mathrm{Des}(T)).$$

Given a partition $\lambda = (\lambda_1, \ldots, \lambda_k)$, the *canonical* SYT V_λ is the unique SYT satisfying $sh(V_\lambda) = \lambda$ and $\mathrm{comp}(V_\lambda) = (\lambda_1, \ldots, \lambda_k)$. In V_λ the first row is filled with $1, 2, \ldots, \lambda_1$ and row i for $2 \leqslant i \leqslant \ell(\lambda)$ is filled with

$$x+1, x+2, \ldots, x+\lambda_i$$

where $x = \lambda_1 + \cdots + \lambda_{i-1}$.

Example 2.4.3.

$$T = \begin{array}{|c|c|c|c|}\cline{1-1} 7 \\ \cline{1-2} 6 & 9 \\ \cline{1-2} 2 & 5 \\ \hline 1 & 3 & 4 & 8 \\ \hline\end{array} \qquad V_{(4,2,2,1)} = \begin{array}{|c|c|c|c|}\cline{1-1} 9 \\ \cline{1-2} 7 & 8 \\ \cline{1-2} 5 & 6 \\ \hline 1 & 2 & 3 & 4 \\ \hline\end{array}$$

$$\mathrm{Des}(T) = \{1,4,5,6,8\}$$
$$\mathrm{comp}(T) = (1,3,1,1,2,1)$$
$$w_{col}(T) = 7621\ 953\ 4\ 8$$

Definition 2.4.4. *Young's lattice* $\mathscr{L}_Y$ is the poset consisting of all partitions with the partial order $\subseteq$ of containment of the corresponding diagrams or, equivalently, the partial order in which $\lambda = (\lambda_1, \ldots, \lambda_\ell)$ is covered by

1. $(\lambda_1, \ldots, \lambda_\ell, 1)$, that is, the partition obtained by suffixing a part of size 1 to λ.
2. $(\lambda_1, \ldots, \lambda_k + 1, \ldots, \lambda_\ell)$, provided that $\lambda_i \neq \lambda_k$ for all $i < k$, that is, the partition obtained by adding 1 to a part of λ as long as that part is the leftmost part of that size.

Example 2.4.5. A saturated chain in $\mathscr{L}_Y$ is

$$(3,1,1) \lessdot_Y (3,2,1) \lessdot_Y (4,2,1) \lessdot_Y (4,2,1,1).$$

To any cover relation $\mu \lessdot_Y \lambda$ in $\mathscr{L}_Y$ we can associate the column number $\mathrm{col}(\mu \lessdot_Y \lambda)$ of the cell that is in the diagram λ but not μ. For example, we have $\mathrm{col}((3,1,1) \lessdot_Y (3,2,1)) = 2$ and $\mathrm{col}((4,3) \lessdot_Y (4,3,1)) = 1$. We extend this notion to the *column sequence* of a saturated chain, which is the sequence of column numbers of the successive cover relations in the chain, that is,

$$\mathrm{col}(\lambda^1 \lessdot_Y \cdots \lessdot_Y \lambda^k) = \mathrm{col}(\lambda^1 \lessdot_Y \lambda^2), \mathrm{col}(\lambda^2 \lessdot_Y \lambda^3), \ldots, \mathrm{col}(\lambda^{k-1} \lessdot_Y \lambda^k).$$

For example,

$$\mathrm{col}(\ (3,1,1) \lessdot_Y (3,2,1) \lessdot_Y (4,2,1) \lessdot_Y (4,2,1,1)\ ) \ = \ 2,4,1.$$

Staying with saturated chains, we end this subsection with a well-known bijection between SYTs and saturated chains in $\mathscr{L}_Y$ implicit in [81, 7.10.3 Proposition].

Proposition 2.4.6. *A one-to-one correspondence between saturated chains in $\mathscr{L}_Y$ and SYTs is given by*

$$\lambda^0 \lessdot_Y \lambda^1 \lessdot_Y \lambda^2 \lessdot_Y \cdots \lessdot_Y \lambda^n \ \leftrightarrow \ T$$

where T is the SYT of shape λ^n/λ^0 such that the number i appears in the cell in T that exists in λ^i but not λ^{i-1}.

Example 2.4.7. The saturated chain in $\mathscr{L}_Y$

$$\begin{aligned} \emptyset \lessdot_Y (1) \lessdot_Y (1,1) \lessdot_Y (2,1) \lessdot_Y (3,1) \lessdot_Y (3,2) \\ \lessdot_Y (3,2,1) \lessdot_Y (3,2,1,1) \lessdot_Y (4,2,1,1) \lessdot_Y (4,2,2,1) \end{aligned}$$

corresponds to the following SYT.

7			
6	9		
2	5		
1	3	4	8

2.5 Reverse tableaux

Closely related to the tableaux of the previous section are reverse tableaux, which we introduce now.

Definition 2.5.1. Given a skew shape λ/μ, we define a *semistandard reverse tableau* (abbreviated to *SSRT*) $\check{T}$ of *shape* $sh(\check{T}) = \lambda/\mu$ to be a filling

$$\check{T} : \lambda/\mu \to \mathbb{Z}^+$$

of the cells of λ/μ such that

1. the entries in each row are weakly decreasing when read from left to right
2. the entries in each column are strictly decreasing when read from bottom to top.

A *standard reverse tableau* (abbreviated to *SRT*) is an SSRT in which the filling is a bijection $\check{T} : \lambda/\mu \to [\,|\lambda/\mu|\,]$, that is, each of the numbers $1, 2, \ldots, |\lambda/\mu|$ appears exactly once. Sometimes we will abuse notation and use SSRTs and SRTs to denote the set of all such tableaux.

Example 2.5.2. An SSRT and SRT, respectively, are shown below.

1			
3	3		
•	6	5	
•	•	6	6
•	•	•	7

3			
4	1		
8	5		
9	7	6	2

Exactly as with SSYTs, given an SSRT $\check{T}$, we define the *content* of $\check{T}$, denoted by $\mathrm{cont}(\check{T})$, to be the list of nonnegative integers

$$\mathrm{cont}(\check{T}) = (c_1, c_2, \ldots, c_{max})$$

where c_i is the number of times i appears in $\check{T}$, and *max* is the largest integer appearing in $\check{T}$. Given variables $x_1, x_2, \ldots$, we define the *monomial of* $\check{T}$ to be

$$x^{\check{T}} = x_1^{c_1} x_2^{c_2} \cdots x_{max}^{c_{max}}.$$

Given an SRT $\check{T}$, its *column reading word*, denoted by $w_{col}(\check{T})$, is obtained by listing the entries from the leftmost column in *increasing* order, followed by the entries from the second leftmost column, again in increasing order, and so on.

The *descent set* of an SRT $\check{T}$ of size n, denoted by $\mathrm{Des}(\check{T})$, is the subset of $[n-1]$ consisting of all entries i of $\check{T}$ such that $i+1$ appears in the same column or a column to the right, that is,

$$\mathrm{Des}(\check{T}) = \{i \mid i+1 \text{ appears weakly right of } i\} \subseteq [n-1]$$

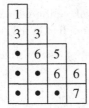

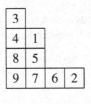

and the corresponding *descent composition* of $\check{T}$ is

$$\mathrm{comp}(\check{T}) = \mathrm{comp}(\mathrm{Des}(\check{T})).$$

Given a partition $\lambda = (\lambda_1, \ldots, \lambda_k) \vdash n$, the *canonical* SRT $\check{V}_\lambda$ is the unique SRT satisfying $sh(\check{V}_\lambda) = \lambda$ and $\mathrm{comp}(\check{V}_\lambda) = (\lambda_k, \ldots, \lambda_1)$. In $\check{V}_\lambda$ the first row is filled with $n, n-1, \ldots, n-\lambda_1+1$ and row i for $2 \leqslant i \leqslant \ell(\lambda)$ is filled with

$$x, x-1, \ldots, x-\lambda_i+1$$

where $x = n - (\lambda_1 + \cdots + \lambda_{i-1})$.

Example 2.5.3.

$$\check{T} = \begin{array}{|c|c|c|c|}\cline{1-1} 3 \\ \cline{1-2} 4 & 1 \\ \cline{1-2} 8 & 5 \\ \hline 9 & 7 & 6 & 2 \\ \hline \end{array} \qquad \check{V}_{(4,2,2,1)} = \begin{array}{|c|c|c|c|}\cline{1-1} 1 \\ \cline{1-2} 3 & 2 \\ \cline{1-2} 5 & 4 \\ \hline 9 & 8 & 7 & 6 \\ \hline \end{array}$$

$$\mathrm{Des}(\check{T}) = \{1,3,4,5,8\}$$
$$\mathrm{comp}(\check{T}) = (1,2,1,1,3,1)$$
$$w_{col}(\check{T}) = 3489\ 157\ 6\ 2$$

There is a natural shape-preserving bijection

$$\check{\Gamma} : SYTs \to SRTs$$

where for an SYT T with n cells we replace each entry i by the entry $n-i+1$, obtaining an SRT $\check{T} = \check{\Gamma}(T)$ of the same skew shape.

Therefore, we have an analogue to Proposition 2.4.6 for SRTs.

Proposition 2.5.4. *A one-to-one correspondence between saturated chains in $\mathscr{L}_Y$ and SRTs is given by*

$$\lambda^0 \lessdot_Y \lambda^1 \lessdot_Y \lambda^2 \lessdot_Y \cdots \lessdot_Y \lambda^n \leftrightarrow \check{T}$$

where $\check{T}$ is the SRT of shape λ^n/λ^0 such that the number $n-i+1$ appears in the cell in $\check{T}$ that exists in λ^i but not λ^{i-1}.

Example 2.5.5. The saturated chain in $\mathscr{L}_Y$

$$\emptyset \lessdot_Y (1) \lessdot_Y (1,1) \lessdot_Y (2,1) \lessdot_Y (3,1) \lessdot_Y (3,2)$$
$$\lessdot_Y (3,2,1) \lessdot_Y (3,2,1,1) \lessdot_Y (4,2,1,1) \lessdot_Y (4,2,2,1)$$

corresponds to the following SRT.

3			
4	1		
8	5		
9	7	6	2

Additionally, there is a simple relationship between descent compositions of SYTs and SRTs. We include its proof as the equivalent statements are useful to know.

Proposition 2.5.6. *Given an SYT T, we have* $\operatorname{comp}(\check{\Gamma}(T))) = \operatorname{comp}(T)^r$.

Proof. Suppose T is an SYT with n cells. The following statements are equivalent.

1. $i \in \operatorname{Des}(T)$.
2. $i+1$ is weakly to the left of i in T.
3. $n-i$ is weakly to the left of $n-i+1$ in $\check{\Gamma}(T)$.
4. $n-i+1$ is weakly to the right of $n-i$ in $\check{\Gamma}(T)$.
5. $n-i \in \operatorname{Des}(\check{\Gamma}(T))$.

This establishes the claim. □

2.6 Schensted insertion

Schensted insertion is an algorithm with many interesting combinatorial properties and applications to representation theory. For further details see [31,72,81]. We will also use this algorithm and the variation below in Chapter 5.

In particular, *Schensted insertion* inserts a positive integer k_1 into a semistandard or standard Young tableau T and is denoted by $T \leftarrow k_1$.

1. If k_1 is greater than or equal to the last entry in row 1, place it at the end of the row, else
2. find the leftmost entry in that row strictly larger than k_1, say k_2, then
3. replace k_2 by k_1, that is, k_1 *bumps* k_2.
4. Repeat the previous steps with k_2 and row 2, k_3 and row 3, etc.

The set of cells whose values are modified by the insertion, including the final cell added, is called the *insertion path*, and the final cell is called the *new cell*.

Example 2.6.1. If we insert 5, then we have

7	7		
5	6	6	
2	4	5	
1	3	4	6

$\leftarrow 5 \quad =$

7	7		
5	6	6	
2	4	5	**6**
1	3	4	**5**

where the bold cells indicate the insertion path. Meanwhile, if we insert 3, then we have

7	7		
5	6	6	
2	4	5	
1	3	4	6

$\leftarrow 3 \quad =$

7			
6	7		
5	**5**	6	
2	4	**4**	
1	3	**3**	6

where the bold cells again indicate the insertion path.

Similarly we have *Schensted insertion for reverse tableaux* [40], which inserts a positive integer k_1 into a semistandard or standard reverse tableau $\check{T}$ and is denoted by $\check{T} \leftarrow k_1$.

1. If k_1 is less than or equal to the last entry in row 1, place it at the end of the row, else
2. find the leftmost entry in that row strictly smaller than k_1, say k_2, then
3. replace k_2 by k_1, that is, k_1 *bumps* k_2.
4. Repeat the previous steps with k_2 and row 2, k_3 and row 3, etc.

As before, the set of cells whose values are modified by the insertion, including the final cell added, is called the *insertion path*, and the final cell is called the *new cell*.

Example 2.6.2. If we insert 5, then we have

1	1		
3	2	2	
6	4	3	
7	5	4	2

$\leftarrow 5 \quad =$

1			
2	1		
3	**3**	2	
6	4	**4**	
7	5	**5**	2

where the bold cells indicate the insertion path.

Given a type of insertion and list of positive integers $\sigma = \sigma_1 \cdots \sigma_n$, we define the P-*tableau*, or *insertion tableau*, or *rectification* of σ, denoted by $\mathrm{P}(\sigma)$, to be

$$(\cdots((\emptyset \leftarrow \sigma_1) \leftarrow \sigma_2) \cdots) \leftarrow \sigma_n.$$

Chapter 3
Hopf algebras

Abstract We give the basic theory of graded Hopf algebras, and then illustrate the theory in detail with three examples: the Hopf algebra of symmetric functions, Sym, the Hopf algebra of quasisymmetric functions, QSym, and the Hopf algebra of noncommutative symmetric functions, NSym. In each case we describe pertinent bases, the product, the coproduct and the antipode. Once defined we see how Sym is a subalgebra of QSym, and a quotient of NSym. We also discuss the duality of QSym and NSym and a variety of automorphisms on each. We end by defining combinatorial Hopf algebras and discussing the role QSym plays as the terminal object in the category of all combinatorial Hopf algebras.

3.1 Hopf algebra basic theory

Here we present all the definitions and results concerning Hopf algebras that we shall need. The reader may wish to use this section as a reference, to be consulted after seeing three examples of a Hopf algebra in Sections 3.2, 3.3 and 3.4. Our presentation is based on that in [69]. Other references are [68] and [84].

Throughout this section, let R be a commutative ring with identity element. We remind the reader that an R-module is defined in the same way as a vector space, except that the field of scalars is replaced by the ring R.

Definition 3.1.1. An *algebra* over R is an R-module $\mathscr{A}$ together with R-linear maps *product* or *multiplication* $m : \mathscr{A} \otimes \mathscr{A} \to \mathscr{A}$ and *unit* $u : R \to \mathscr{A}$, such that the following diagrams commute.

K. Luoto et al., *An Introduction to Quasisymmetric Schur Functions*,
SpringerBriefs in Mathematics, DOI 10.1007/978-1-4614-7300-8_3,
© Kurt Luoto, Stefan Mykytiuk, Stephanie van Willigenburg 2013

i. *Associative property*

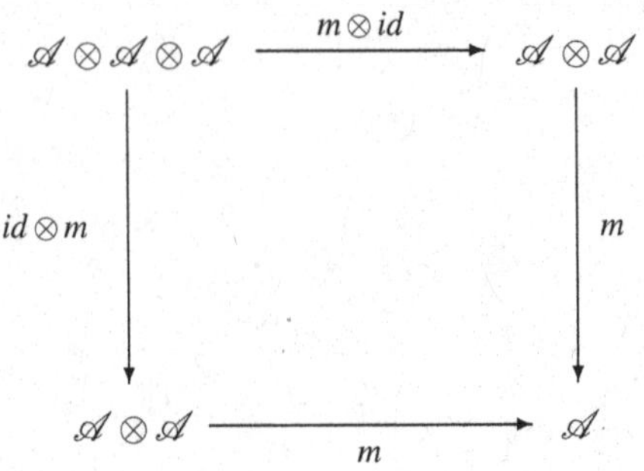

ii. *Unitary property*

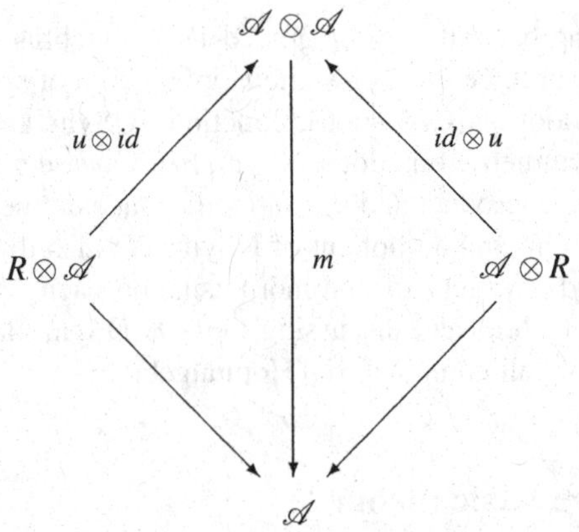

In both diagrams, *id* is the identity map on $\mathscr{A}$. The two lower maps in diagram ii. are given by scalar multiplication.

A map $f : \mathscr{A} \to \mathscr{A}'$, where $(\mathscr{A}', m', u')$ is another algebra over R, is an *algebra morphism* if

$$f \circ m = m' \circ (f \otimes f) \text{ and } f \circ u = u'.$$

We shall frequently write ab instead of $m(a \otimes b)$.

The algebra $\mathscr{A}$ has identity element $1_{\mathscr{A}} = u(1_R)$, where 1_R is the identity element of R. The unit u is always given by $u(r) = r1_{\mathscr{A}}$ for all $r \in R$.

A coalgebra is defined by reversing the arrows in the diagrams that define an algebra.

Definition 3.1.2. A *coalgebra* over R is an R-module $\mathscr{C}$ together with R-linear maps *coproduct* or *comultiplication* $\Delta : \mathscr{C} \to \mathscr{C} \otimes \mathscr{C}$ and *counit* or *augmentation* $\varepsilon : \mathscr{C} \to R$, such that the following diagrams commute.

i. *Coassociativity property*

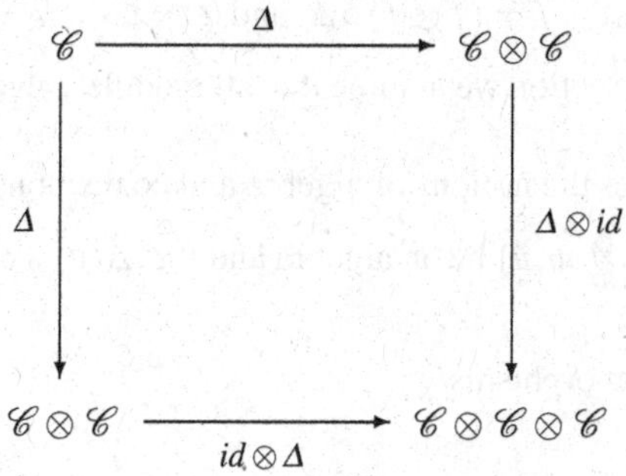

ii. *Counitary property*

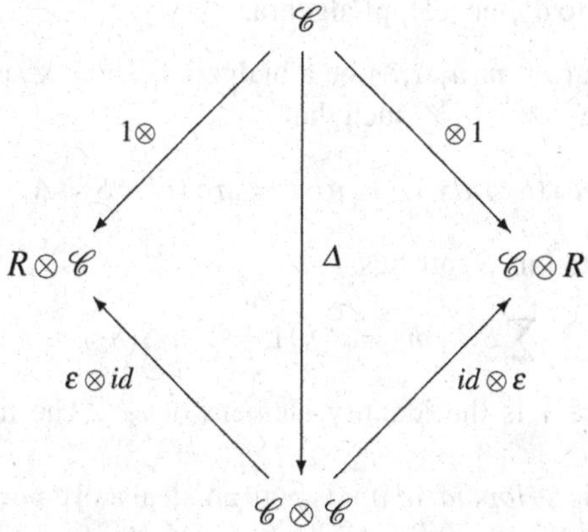

In both diagrams, *id* is the identity map on $\mathscr{C}$. The two upper maps in diagram ii. are given by $(1\otimes)(c) = 1 \otimes c$ and $(\otimes 1)(c) = c \otimes 1$ for $c \in C$. We may omit the indexing required to express a coproduct $\Delta(c)$ as an element of $\mathscr{C} \otimes \mathscr{C}$ and use *Sweedler notation* to write $\Delta(c) = \sum c_1 \otimes c_2$. Thus, in Sweedler notation, diagrams i. and ii. (together with bilinearity of $\otimes$) state that, for all $c \in C$,

$$\sum c_1 \otimes (c_2)_1 \otimes (c_2)_2 = \sum (c_1)_1 \otimes (c_1)_2 \otimes c_2 \tag{3.1}$$

and

$$\sum \varepsilon(c_1) c_2 = c = \sum \varepsilon(c_2) c_1. \tag{3.2}$$

We say the coproduct Δ is *cocommutative* if $c'' \otimes c'$ is a term of $\Delta(c)$ whenever $c' \otimes c''$ is.

A submodule $\mathscr{I} \subseteq \mathscr{C}$ is a *coideal* if

$$\Delta(\mathscr{I}) \subseteq \mathscr{I} \otimes \mathscr{C} + \mathscr{C} \otimes \mathscr{I} \text{ and } \varepsilon(\mathscr{I}) = \{0\}.$$

A map $f : \mathscr{C} \to \mathscr{C}'$, where $(\mathscr{C}', \Delta', \varepsilon')$ is another coalgebra over R, is a *coalgebra morphism* if

$$\Delta' \circ f = (f \otimes f) \circ \Delta \text{ and } \varepsilon = \varepsilon' \circ f.$$

In the remainder of this section we assume that all modules, algebras and coalgebras are over the ring R.

A bialgebra combines the notions of algebra and coalgebra.

Definition 3.1.3. Let $(\mathscr{B}, m, u)$ be an algebra and $(\mathscr{B}, \Delta, \varepsilon)$ a coalgebra. Then $\mathscr{B}$ is a *bialgebra* if

1. Δ and ε are algebra morphisms

or equivalently,

2. m and u are coalgebra morphisms.

We are now ready to define a Hopf algebra.

Definition 3.1.4. Let $(\mathscr{H}, m, u, \Delta, \varepsilon)$ be a bialgebra. Then $\mathscr{H}$ is a *Hopf algebra* if there is a linear map $S : \mathscr{H} \to \mathscr{H}$ such that

$$m \circ (S \otimes id) \circ \Delta = u \circ \varepsilon = m \circ (id \otimes S) \circ \Delta.$$

Thus, in Sweedler notation, S satisfies

$$\sum S(h_1)h_2 = \varepsilon(h)1 = \sum h_1 S(h_2)$$

for all $h \in \mathscr{H}$, where 1 is the identity element of $\mathscr{H}$. The map S is called the *antipode* of $\mathscr{H}$.

A subset $\mathscr{I} \subseteq \mathscr{H}$ is a *Hopf ideal* if it is both an ideal and coideal, and $S(\mathscr{I}) \subseteq \mathscr{I}$.

A map $f : \mathscr{H} \to \mathscr{H}'$ between Hopf algebras is a *Hopf morphism* if it is both an algebra and coalgebra morphism, and

$$f \circ S_{\mathscr{H}} = S_{\mathscr{H}'} \circ f$$

where $S_{\mathscr{H}}$ and $S_{\mathscr{H}'}$ are respectively the antipodes of $\mathscr{H}$ and $\mathscr{H}'$.

We note that the antipode of a Hopf algebra is unique. The following result is useful in proving that algebras are Hopf.

Proposition 3.1.5. *Let $\mathscr{I}$ be a submodule of a Hopf algebra $\mathscr{H}$. Then $\mathscr{I}$ is a Hopf ideal if and only if $\mathscr{H}/\mathscr{I}$ is a Hopf algebra with structure induced by $\mathscr{H}$.*

The existence of an antipode is guaranteed in the following type of bialgebra.

Definition 3.1.6. A bialgebra $\mathscr{B}$ with coproduct Δ is *graded* if it contains submodules $\mathscr{B}_0, \mathscr{B}_1, \ldots$ satisfying

1. $\mathscr{B} = \bigoplus_{n \geqslant 0} \mathscr{B}^n$,
2. $\mathscr{B}^i \mathscr{B}^j \subseteq \mathscr{B}^{i+j}$,
3. $\Delta(\mathscr{B}^n) \subseteq \bigoplus_{i+j=n} \mathscr{B}^i \otimes \mathscr{B}^j$.

Elements of the submodule $\mathscr{B}^n$ are called *homogeneous of degree n*. If $\mathscr{B}^0$ has dimension 1, we say $\mathscr{B}$ is *connected.*

The following result is a consequence of [68, Proposition 8.2].

Proposition 3.1.7. *[27, Lemma 2.1] Let $\mathscr{B}$ be a connected graded bialgebra. Then $\mathscr{B}$ is a Hopf algebra with unique antipode S defined recursively by $S(1) = 1$, and for x of degree $n \geqslant 1$,*

$$S(x) = -\sum_{i=0}^{n-1} S(y_i)z_{n-i},$$

where

$$\Delta(x) = x \otimes 1 + \sum_{i=0}^{n-1} y_i \otimes z_{n-i} \tag{3.3}$$

and y_i, z_i have degree i.

The counitary property ensures that the coproduct of an element x of degree n in a connected graded bialgebra has the expansion shown in Equation (3.3).

The Hopf algebras that we shall study are infinite-dimensional, graded and connected, with each component of the direct sum having finite dimension. Associated with each such Hopf algebra is another Hopf algebra of interest to us. This other Hopf algebra is described in the following, which is a consequence of [69, Theorem 9.1.3].

Proposition 3.1.8. *Let $\mathscr{H} = \bigoplus_{n \geqslant 0} \mathscr{H}^n$ be a connected, graded Hopf algebra over R, such that each homogeneous component $\mathscr{H}^n$ is finite-dimensional. Define the module $\mathscr{H}^*$ by*

$$\mathscr{H}^* = \bigoplus_{n \geqslant 0} (\mathscr{H}^n)^*,$$

where $(\mathscr{H}^n)^*$ denotes the set of all linear maps $f : \mathscr{H}^n \to R$.

Then $\mathscr{H}^*$ is a Hopf algebra with

1. *product $m : \mathscr{H}^* \otimes \mathscr{H}^* \to \mathscr{H}^*$ induced by the* convolution product

$$f * g = m_R \circ (f \otimes g) \circ \Delta_{\mathscr{H}},$$

where m_R is the product of R, and $\Delta_{\mathscr{H}}$ is the coproduct of $\mathscr{H}$ (in Sweedler notation, the convolution product is given by

$$(f * g)(h) = \sum f(h_1)g(h_2)$$

for all $h \in \mathscr{H}$),

2. *unit $u : R \to \mathscr{H}^*$ given by*

$$u(r) = r\varepsilon_{\mathscr{H}},$$

where $\varepsilon_{\mathscr{H}}$ is the counit of $\mathscr{H}$ (thus $\varepsilon_{\mathscr{H}}$ is the identity element of $\mathscr{H}^*$),

3. *coproduct* $\Delta : \mathcal{H}^* \to \mathcal{H}^* \otimes \mathcal{H}^*$ *given by*

$$\Delta(f) = \rho^{-1}(f \circ m_{\mathcal{H}}),$$

where ρ *is the invertible linear map defined on* $\mathcal{H}^* \otimes \mathcal{H}^*$ *by*

$$(\rho(f \otimes g))(h_1 \otimes h_2) = f(h_1)g(h_2)$$

and $m_{\mathcal{H}}$ *is the product of* $\mathcal{H}$,

4. *counit* $\varepsilon : \mathcal{H}^* \to R$ *given by*

$$\varepsilon(f) = f(1_{\mathcal{H}}),$$

where $1_{\mathcal{H}}$ *is the identity element of* $\mathcal{H}$,

5. *antipode* $S : \mathcal{H}^* \to \mathcal{H}^*$ *given by*

$$S(f) = f \circ S_{\mathcal{H}},$$

where $S_{\mathcal{H}}$ *is the antipode of* $\mathcal{H}$.

The Hopf algebra $\mathcal{H}^*$ *is called the* graded Hopf dual *of* $\mathcal{H}$.

Accordingly, there is a nondegenerate bilinear form $\langle \cdot, \cdot \rangle : \mathcal{H} \otimes \mathcal{H}^* \to R$ that pairs the elements of any basis $\{B_i\}_{i \in I}$ of $\mathcal{H}^n$ for some index set I, and its dual basis $\{D_i\}_{i \in I}$ of $(\mathcal{H}^n)^*$, given by $\langle B_i, D_j \rangle = \delta_{ij}$, where the *Kronecker delta* $\delta_{ij} = 1$ if $i = j$ and 0 otherwise. Duality is exhibited in that the product coefficients of one basis are the coproduct coefficients of its dual basis and vice versa, that is,

$$B_i B_j = \sum_h a_{i,j}^h B_h \quad \Longleftrightarrow \quad \Delta(D_h) = \sum_{i,j} a_{i,j}^h D_i \otimes D_j,$$

$$D_i D_j = \sum_h b_{i,j}^h D_h \quad \Longleftrightarrow \quad \Delta(B_h) = \sum_{i,j} b_{i,j}^h B_i \otimes B_j.$$

3.2 The Hopf algebra of symmetric functions

We now introduce the Hopf algebra of symmetric functions. A more extensive treatment can be found in the books [60,72,81] but here we restrict ourselves in order to illuminate certain parallels with quasisymmetric functions and noncommutative symmetric functions later.

Let $\mathbb{Q}[[x_1, x_2, \ldots]]$ be the Hopf algebra of formal power series in infinitely many variables $x_1, x_2, \ldots$ over $\mathbb{Q}$. Given a monomial $x_{i_1}^{\alpha_1} \cdots x_{i_k}^{\alpha_k}$, we say it has *degree* n if $(\alpha_1, \ldots, \alpha_k) \vDash n$. Furthermore we say a formal power series has *finite degree* if each monomial has degree at most m for some nonnegative integer m, and is *homogeneous of degree* n if each monomial has degree n.

Definition 3.2.1. A *symmetric function* is a formal power series $f \in \mathbb{Q}[[x_1, x_2, \ldots]]$ such that

1. The degree of f is finite.
2. For every composition $(\alpha_1, \ldots, \alpha_k)$, all monomials $x_{i_1}^{\alpha_1} \cdots x_{i_k}^{\alpha_k}$ in f with distinct indices $i_1, \ldots, i_k$ have the same coefficient.

Let n be a nonnegative integer, then recall that a *permutation* of $[n]$ is a bijection $\sigma : [n] \to [n]$, which we may write as an n-tuple $\sigma(1) \cdots \sigma(n)$. The set of all permutations of $[n]$ is denoted by $\mathfrak{S}_n$, and the union $\bigcup_{n \geqslant 0} \mathfrak{S}_n$ by $\mathfrak{S}_\infty$. We identify a permutation $\sigma \in \mathfrak{S}_n$ with a bijection of the positive integers by defining $\sigma(i) = i$ if $i > n$. Then $\mathfrak{S}_0 \subset \mathfrak{S}_1 \subset \cdots$ and $\mathfrak{S}_\infty$ becomes a group, known as a *symmetric group*, with the operation of map composition. The identity element is the unique permutation of $\emptyset$. Given $\sigma = \sigma_1 \cdots \sigma_n \in \mathfrak{S}_n$, we define its *descent set*, denoted by $d(\sigma)$, to be

$$d(\sigma) = \{i \mid \sigma(i) > \sigma(i+1)\} \subseteq [n-1].$$

Equivalently we can think of a symmetric function as follows.

Definition 3.2.2. A *symmetric function* is a formal power series $f \in \mathbb{Q}[[x_1, x_2, \ldots]]$ such that

1. f has finite degree,
2. f is invariant under the action of $\mathfrak{S}_\infty$ on $\mathbb{Q}[[x_1, x_2, \ldots]]$ given by

$$\sigma.(x_{i_1}^{\alpha_1} \cdots x_{i_k}^{\alpha_k}) = x_{\sigma(i_1)}^{\alpha_1} \cdots x_{\sigma(i_k)}^{\alpha_k}.$$

That is, $\sigma.f = f$ when the action of σ is extended by linearity.

Example 3.2.3. If $f = x_1^2 + x_2^2 + x_1 x_2$ and $\sigma \in \mathfrak{S}_2$, then

$$\sigma.f = x_1^2 + x_2^2 + x_1 x_2 = f.$$

The set of all symmetric functions with the operations of the next subsection forms a graded Hopf algebra

$$\mathrm{Sym} = \bigoplus_{n \geqslant 0} \mathrm{Sym}^n$$

spanned by the following functions, strongly suggested by the definition of Sym.

Definition 3.2.4. Let $\lambda = (\lambda_1, \ldots, \lambda_k)$ be a partition. Then the *monomial symmetric function* m_λ is defined by

$$m_\lambda = \sum x_{i_1}^{\lambda_1} \cdots x_{i_k}^{\lambda_k},$$

where the sum is over all k-tuples $(i_1, \ldots, i_k)$ of distinct indices that yield distinct monomials. We define $m_\emptyset = 1$.

Example 3.2.5. We have

$$m_{(2,1)} = x_1^2x_2^1 + x_2^2x_1^1 + x_1^2x_3^1 + x_3^2x_1^1 + x_1^2x_4^1 + x_4^2x_1^1 + x_2^2x_3^1 + x_3^2x_2^1 + \cdots.$$

Moreover, since the m_λ are independent we have

$$\mathrm{Sym}^n = \mathrm{span}\{m_\lambda \mid \lambda \vdash n\}.$$

The basis of monomial symmetric functions is not the only interesting and useful basis.

Definition 3.2.6. Let n be a nonnegative integer. Then the *n-th elementary symmetric function*, denoted by e_n, is defined by

$$e_n = m_{(1^n)} = \sum_{i_1 < \cdots < i_n} x_{i_1} \cdots x_{i_n}$$

and the *n-th complete homogeneous symmetric function*, denoted by h_n, is defined by

$$h_n = \sum_{\lambda \vdash n} m_\lambda = \sum_{i_1 \leqslant \cdots \leqslant i_n} x_{i_1} \cdots x_{i_n}$$

with $e_0 = h_0 = 1$.

Let $\lambda = (\lambda_1, \ldots, \lambda_k)$ be a partition. Then the *elementary symmetric function* e_λ is defined by

$$e_\lambda = e_{\lambda_1} \cdots e_{\lambda_k} = \prod_{\lambda_i} e_{\lambda_i}$$

and the *complete homogeneous symmetric function* h_λ is defined by

$$h_\lambda = h_{\lambda_1} \cdots h_{\lambda_k} = \prod_{\lambda_i} h_{\lambda_i} = \sum_{\beta \preccurlyeq (\lambda_1, \ldots, \lambda_k)} (-1)^{|\lambda| - \ell(\beta)} e_{\widetilde{\beta}}.$$

We have, in particular,

$$h_n = \sum_{\beta \vDash n} (-1)^{n - \ell(\beta)} e_{\widetilde{\beta}}.$$

Example 3.2.7. Note that $h_1 = e_1 = m_{(1)} = x_1 + x_2 + x_3 + \cdots$ while

$$h_2 = m_{(2)} + m_{(1,1)} = x_1^2 + x_2^2 + x_3^2 + \cdots + x_1x_2 + x_1x_3 + x_2x_3 + \cdots$$

and

$$e_2 = m_{(1,1)} = x_1x_2 + x_1x_3 + x_2x_3 + \cdots.$$

Hence

$$h_{(2,1)} = h_2h_1 = (x_1^2 + x_2^2 + x_3^2 + \cdots + x_1x_2 + x_1x_3 + x_2x_3 + \cdots)(x_1 + x_2 + x_3 + \cdots)$$

$$e_{(2,1)} = e_2 e_1 = (x_1 x_2 + x_1 x_3 + x_2 x_3 + \cdots)(x_1 + x_2 + x_3 + \cdots).$$

These bases are also of interest. For example, the *fundamental theorem of symmetric functions* states that Sym is a polynomial algebra in the elementary symmetric functions, that is

$$\text{Sym} = \mathbb{Q}[e_1, e_2, \ldots].$$

However, the most important basis of Sym is considered to be the basis of Schur functions, due to its connections to other areas of mathematics such as representation theory and algebraic geometry. Unlike the preceding bases, it is not obvious that these basis elements are symmetric.

Definition 3.2.8. Let λ be a partition. Then the *Schur function* s_λ is defined to be $s_\emptyset = 1$ and

$$s_\lambda = \sum_T x^T$$

where the sum is over all SSYTs (or equivalently SSRTs) T of shape λ.

Example 3.2.9. We have $s_{(2,1)} = x_1^2 x_2 + x_2^2 x_1 + \ldots + 2x_1 x_2 x_3 + \cdots$ from the SSYTs

$$
\begin{array}{cc}
\boxed{2} & \\
\boxed{1}\boxed{1} &
\end{array}
\qquad
\begin{array}{cc}
\boxed{2} & \\
\boxed{1}\boxed{2} &
\end{array}
\qquad
\begin{array}{cc}
\boxed{3} & \\
\boxed{1}\boxed{2} &
\end{array}
\qquad
\begin{array}{cc}
\boxed{2} & \\
\boxed{1}\boxed{3} &
\end{array}
$$

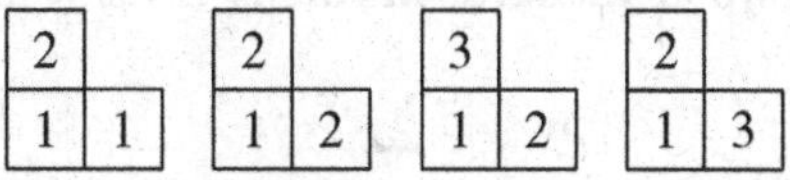

or equivalently from the following SSRTs.

$$
\begin{array}{cc}
\boxed{1} & \\
\boxed{2}\boxed{1} &
\end{array}
\qquad
\begin{array}{cc}
\boxed{1} & \\
\boxed{2}\boxed{2} &
\end{array}
\qquad
\begin{array}{cc}
\boxed{1} & \\
\boxed{3}\boxed{2} &
\end{array}
\qquad
\begin{array}{cc}
\boxed{2} & \\
\boxed{3}\boxed{1} &
\end{array}
$$

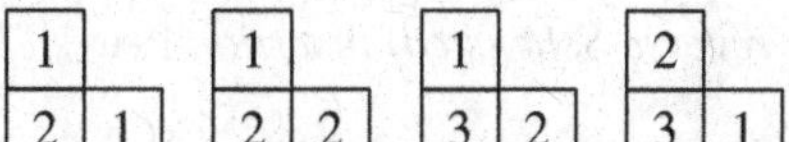

Schur functions can also be expressed in terms of SSYTs or SSRTs when expanded in the basis of monomial symmetric functions.

Proposition 3.2.10. *Let* $\lambda \vdash n$. *Then*

$$s_\lambda = \sum_{\mu \vdash n} K_{\lambda\mu} m_\mu$$

where $s_\emptyset = 1$ *and* $K_{\lambda\mu}$ *is the number of SSYTs (or equivalently SSRTs) T satisfying* $\text{sh}(T) = \lambda$ *and* $\text{cont}(T) = \mu$. *The* $K_{\lambda\mu}$ *are known as* Kostka numbers.

Example 3.2.11. We have $s_{(2,1)} = m_{(2,1)} + 2m_{(1,1,1)}$ from the SSYTs

$$
\begin{array}{cc}
\boxed{2} & \\
\boxed{1}\boxed{1} &
\end{array}
\qquad
\begin{array}{cc}
\boxed{3} & \\
\boxed{1}\boxed{2} &
\end{array}
\qquad
\begin{array}{cc}
\boxed{2} & \\
\boxed{1}\boxed{3} &
\end{array}
$$

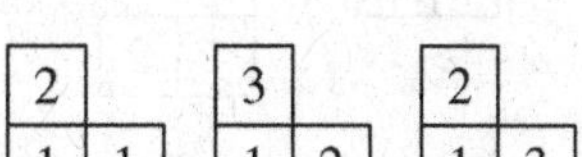

or equivalently from the following SSRTs.

$$\begin{array}{|c|c}\hline 1 & \\ \hline 2 & \multicolumn{1}{c|}{1} \\ \hline \end{array} \qquad \begin{array}{|c|c}\hline 1 & \\ \hline 3 & \multicolumn{1}{c|}{2} \\ \hline \end{array} \qquad \begin{array}{|c|c}\hline 2 & \\ \hline 3 & \multicolumn{1}{c|}{1} \\ \hline \end{array}$$

Note that if in Definition 3.2.8 the straight shape λ was replaced by a skew shape λ/μ, then a function would still be defined. These functions also play a role in the theory of symmetric functions.

Definition 3.2.12. Let λ/μ be a skew shape. Then the *skew Schur function* $s_{\lambda/\mu}$ is defined to be $s_\emptyset = 1$ and

$$s_{\lambda/\mu} = \sum_T x^T$$

where the sum is over all SSYTs (or equivalently SSRTs) T of shape λ/μ.

The expansion of a skew Schur function as a linear combination of Schur functions is the celebrated Littlewood-Richardson rule, first conjectured in [56] and proved much later in [73, 86]. Many versions of it exist, and a number of these can be found in [81, Appendix A1.3] or [31, Chapter 5], one of which we now give.

Theorem 3.2.13 (Littlewood-Richardson rule). *Let μ, ν be partitions. Then*

$$s_{\nu/\mu} = \sum c^{\nu}_{\lambda\mu} s_\lambda$$

where the sum is over all partitions λ, and the Littlewood-Richardson coefficient *$c^{\nu}_{\lambda\mu}$ counts the number of SYTs (respectively SRTs) T of shape ν/μ such that using Schensted (respectively reverse Schensted) insertion* $\mathrm{P}(w_{col}(T)) = V_\lambda$ *(respectively $\check{V}_\lambda$).*

Example 3.2.14. We have $s_{(2,2,1)/(1)} = s_{(2,2)} + s_{(2,1,1)}$ from the SYTs

$$\begin{array}{|c|c|}\cline{1-1} 3 & \multicolumn{1}{c}{} \\ \hline 1 & 4 \\ \hline \multicolumn{1}{c|}{} & 2 \\ \cline{2-2} \end{array} \qquad \begin{array}{|c|c|}\cline{1-1} 4 & \multicolumn{1}{c}{} \\ \hline 1 & 3 \\ \hline \multicolumn{1}{c|}{} & 2 \\ \cline{2-2} \end{array}$$

with respective column reading words 3142 and 4132 whose respective P-tableaux are the following canonical SYTs.

$$\begin{array}{|c|c|}\hline 3 & 4 \\ \hline 1 & 2 \\ \hline \end{array} \qquad \begin{array}{|c|c}\hline 4 & \\ \cline{1-1} 3 & \\ \hline 1 & \multicolumn{1}{c|}{2} \\ \hline \end{array}$$

When a skew Schur function is indexed by a ribbon we call it a *ribbon Schur function*, denoted by r_α where α is the composition corresponding to the ribbon.

Ribbon Schur functions have a particularly appealing expansion in terms of the complete homogeneous symmetric functions, which goes back to MacMahon [61].

Proposition 3.2.15. *For any composition α*

$$r_\alpha = \sum_{\beta \succcurlyeq \alpha} (-1)^{\ell(\alpha)-\ell(\beta)} h_{\widetilde{\beta}}.$$

As an example $r_{(1,2,1)} = h_{(2,1,1)} - 2h_{(3,1)} + h_{(4)}$. In fact, another basis for Sym^n is given by $\{r_\lambda \mid \lambda \vdash n\}$ since by Proposition 3.2.15 the matrix expressing ribbon Schur functions in terms of complete homogeneous symmetric functions is upper triangular with ± 1 on the diagonal if we index the rows and columns using a linear extension of refinement. This result also follows from [18, Proposition 2.2]. To summarize our bases we have

$$\begin{aligned}\mathrm{Sym}^n &= \mathrm{span}\{m_\lambda \mid \lambda \vdash n\} = \mathrm{span}\{e_\lambda \mid \lambda \vdash n\} = \mathrm{span}\{h_\lambda \mid \lambda \vdash n\} \\ &= \mathrm{span}\{s_\lambda \mid \lambda \vdash n\} = \mathrm{span}\{r_\lambda \mid \lambda \vdash n\}.\end{aligned}$$

3.2.1 Products and coproducts

There is no simple description for the product of two monomial symmetric functions, but this is not the case for the other bases introduced in the previous section. From their definitions it is immediate that

$$e_\lambda e_\mu = e_{\widetilde{\lambda \cdot \mu}} \tag{3.4}$$

$$h_\lambda h_\mu = h_{\widetilde{\lambda \cdot \mu}}. \tag{3.5}$$

For example, $h_{(4,2,1,1)} h_{(3,1,1)} = h_{(4,3,2,1,1,1,1)}$. Concerning the spanning set of all ribbon Schur functions we have that for compositions α and β

$$r_\alpha r_\beta = r_{\alpha \cdot \beta} + r_{\alpha \odot \beta}. \tag{3.6}$$

For example,

$$r_{(1,4,1,2)} r_{(3,1,1)} = r_{(1,4,1,2,3,1,1)} + r_{(1,4,1,5,1,1)}.$$

However, the product of two Schur functions expanded as a linear combination of Schur functions is another incarnation of the Littlewood-Richardson rule in Theorem 3.2.13.

Theorem 3.2.16 (Littlewood-Richardson rule). *Let λ, μ be partitions. Then*

$$s_\lambda s_\mu = \sum c^\nu_{\lambda\mu} s_\nu$$

where the sum is over all partitions ν, *and the* Littlewood-Richardson coefficient $c^{\nu}_{\lambda\mu}$ *counts the number of SYTs (respectively SRTs) T of shape* ν/μ *such that using Schensted (respectively reverse Schensted) insertion* $\mathrm{P}(w_{col}(T)) = V_\lambda$ *(respectively* $\check{V}_\lambda$*).*

Special cases of this rule when $s_\lambda = s_{(n)} = h_n$ and $s_\lambda = s_{(1^n)} = e_n$ have simpler descriptions, and are known as the Pieri rules.

Theorem 3.2.17 (Pieri rules). *Let* μ *be a partition and n a nonnegative integer. Then*

$$s_{(n)}s_\mu = h_n s_\mu = \sum s_\nu$$

where the sum is over all partitions ν, *such that* ν/μ *is a horizontal strip with n cells. Similarly,*

$$s_{(1^n)}s_\mu = e_n s_\mu = \sum s_\nu$$

where the sum is over all partitions ν, *such that* ν/μ *is a vertical strip with n cells.*

Example 3.2.18. If $\mu = (2,2,1)$ and $n = 2$, then

$$h_2 s_{(2,2,1)} = s_{(4,2,1)} + s_{(3,2,2)} + s_{(3,2,1,1)} + s_{(2,2,2,1)}$$

from the following additions of two cells to $(2,2,1)$ that form a row strip.

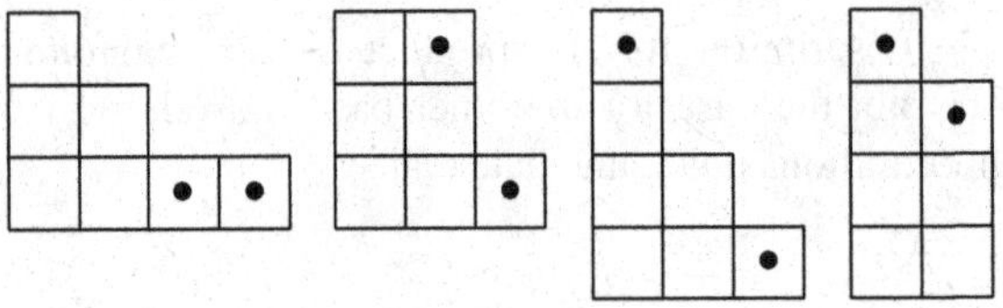

Meanwhile,

$$e_2 s_{(2,2,1)} = s_{(3,2,2)} + s_{(3,2,1,1)} + s_{(2,2,2,1)} + s_{(3,3,1)} + s_{(2,2,1,1,1)}$$

from the last three diagrams above and the two below.

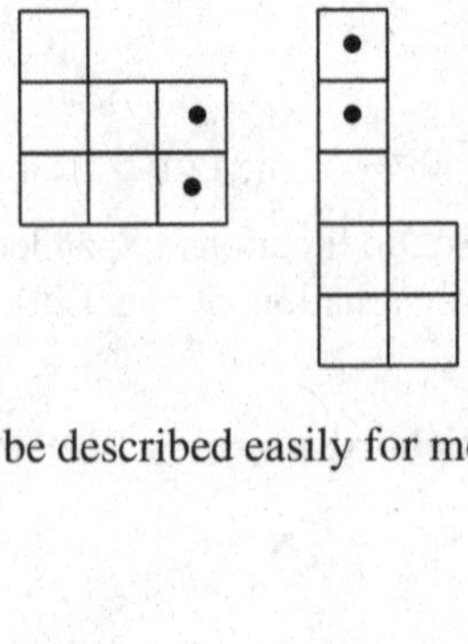

The coproduct can also be described easily for most of our bases of Sym.

$$\Delta(m_\lambda) = \sum_{\lambda=\widetilde{\mu\cdot\nu}} m_\mu \otimes m_\nu \quad \Delta(e_n) = \sum_{i=0}^{n} e_i \otimes e_{n-i} \quad \Delta(h_n) = \sum_{i=0}^{n} h_i \otimes h_{n-i}$$

$$\Delta(s_\lambda) = \sum_{\mu\subseteq\lambda} s_{\lambda/\mu} \otimes s_\mu = \sum_{\mu}\sum_{\nu} c^{\lambda}_{\nu\mu} s_\nu \otimes s_\mu.$$

Example 3.2.19. For the monomial symmetric function $m_{(2,2,1)}$ we have

$$\begin{aligned}\Delta(m_{(2,2,1)}) =& m_{(2,2,1)} \otimes 1 + m_{(2,1)} \otimes m_{(2)} + m_{(2,2)} \otimes m_{(1)} \\ &+ m_{(2)} \otimes m_{(2,1)} + m_{(1)} \otimes m_{(2,2)} + 1 \otimes m_{(2,2,1)}\end{aligned}$$

while for $n=3$ we have

$$\Delta(e_3) = 1 \otimes e_3 + e_1 \otimes e_2 + e_2 \otimes e_1 + e_3 \otimes 1$$

$$\Delta(h_3) = 1 \otimes h_3 + h_1 \otimes h_2 + h_2 \otimes h_1 + h_3 \otimes 1$$

and the coproduct of the Schur function $s_{(2,1)}$ is

$$\Delta(s_{(2,1)}) = s_{(2,1)} \otimes 1 + s_{(2,1)/(1)} \otimes s_{(1)} + s_{(2,1)/(2)} \otimes s_{(2)} + s_{(2,1)/(1,1)} \otimes s_{(1,1)} + 1 \otimes s_{(2,1)}.$$

The counit of Sym is given by

$$\varepsilon(m_\lambda) = \begin{cases} 1 \text{ if } \lambda = \emptyset \\ 0 \text{ otherwise} \end{cases}$$

and an explicit formula for the antipode of a Schur function indexed by a partition λ of n is given by

$$S(s_\lambda) = (-1)^n s_{\lambda^t}. \tag{3.7}$$

3.2.2 Duality

We have noted that Sym is a Hopf algebra, and in fact it is self-dual, that is, Sym and Sym* are isomorphic as Hopf algebras. The bilinear form that pairs the elements of any basis of Sym with its dual basis is called the *Hall inner product*. Since the bases of monomial symmetric functions and complete homogeneous symmetric functions are dual bases, the Hall inner product satisfies

$$\langle m_\lambda, h_\mu \rangle = \delta_{\lambda\mu}. \tag{3.8}$$

It transpires that the basis of Schur functions is self-dual and orthonormal, that is,

$$\langle s_\lambda, s_\mu \rangle = \delta_{\lambda\mu} \tag{3.9}$$

and hence by the Littlewood-Richardson rule and duality we have for partitions λ, μ, ν that

$$\langle s_{\nu/\mu}, s_\lambda \rangle = \langle s_\nu, s_\lambda s_\mu \rangle = c^\nu_{\lambda\mu}. \tag{3.10}$$

3.3 The Hopf algebra of quasisymmetric functions

We now introduce quasisymmetric functions, originally defined by Gessel [35], before connecting them to symmetric functions.

Definition 3.3.1. We say a *quasisymmetric function* is a formal power series $f \in \mathbb{Q}[[x_1, x_2, \ldots]]$ such that

1. The degree of f is finite.
2. For every composition $(\alpha_1, \ldots, \alpha_k)$, all monomials $x_{i_1}^{\alpha_1} \cdots x_{i_k}^{\alpha_k}$ in f with indices $i_1 < \cdots < i_k$ have the same coefficient.

We denote the set of all quasisymmetric functions by QSym.

This definition is analogous to Definition 3.2.1 for symmetric functions. Therefore, one might hope that quasisymmetric functions can be defined as the invariants under some action of the group $\mathfrak{S}_\infty$ of permutations on formal power series, analogous to Definition 3.2.2 for symmetric functions. Such a definition is due to Hivert [46].

Definition 3.3.2. We shall use the following notation. Given a composition $\alpha = (\alpha_1, \ldots, \alpha_k)$ and a k-tuple $I = (i_1, \ldots, i_k)$ of positive integers $i_1 < \cdots < i_k$, let x_I^α denote the monomial $x_{i_1}^{\alpha_1} \cdots x_{i_k}^{\alpha_k}$.

A *quasisymmetric function* is a formal power series $f \in \mathbb{Q}[[x_1, x_2, \ldots]]$ such that

1. f has finite degree,
2. f is invariant under the action of $\mathfrak{S}_\infty$ on $\mathbb{Q}[[x_1, x_2, \ldots]]$ given by

$$\sigma . x_I^\alpha = x_{\sigma . I}^\alpha,$$

where $\sigma . I$ is defined to be the k-tuple obtained by arranging the numbers $\sigma(i_1), \ldots, \sigma(i_k)$ in increasing order. That is, $\sigma . f = f$ where the action of σ is extended by linearity.

Example 3.3.3. If $f = x_1^2 x_2 + x_1^2 x_3 + x_2^2 x_3$ and $\sigma \in \mathfrak{S}_2$, then

$$\sigma . f = x_1^2 x_2 + x_1^2 x_3 + x_2^2 x_3 = f.$$

Unlike the earlier action, the action just described extends to an action ϕ of the algebra $\mathbb{Q}\mathfrak{S}_\infty$ on $\mathbb{Q}[[x_1, x_2, \ldots]]$ that is not faithful. Hivert showed that the quotient $\mathbb{Q}\mathfrak{S}_\infty / \ker \phi$ is isomorphic to the Temperley-Lieb algebra TL_∞. Thus TL_∞ acts faithfully on $\mathbb{Q}[[x_1, x_2, \ldots]]$ and the set of invariants under this action is QSym.

The set of all quasisymmetric functions with the operations of the next subsection forms a graded Hopf algebra

$$\mathrm{QSym} = \bigoplus_{n \geqslant 0} \mathrm{QSym}^n$$

spanned by the following functions, suggested by the definition of QSym.

Definition 3.3.4. Let $\alpha = (\alpha_1, \dots, \alpha_k)$ be a composition. Then the *monomial quasisymmetric function* M_α is defined by

$$M_\alpha = \sum x_{i_1}^{\alpha_1} \cdots x_{i_k}^{\alpha_k},$$

where the sum is over all k-tuples $(i_1, \dots, i_k)$ of indices $i_1 < \cdots < i_k$. We define $M_\emptyset = 1$.

Example 3.3.5. We have

$$M_{(2,1)} = x_1^2 x_2^1 + x_1^2 x_3^1 + x_1^2 x_4^1 + x_2^2 x_3^1 + \cdots$$

while

$$M_{(1,2)} = x_1^1 x_2^2 + x_1^1 x_3^2 + x_1^1 x_4^2 + x_2^1 x_3^2 + \cdots.$$

Since the M_α are independent we have

$$\mathrm{QSym}^n = \mathrm{span}\{M_\alpha \mid \alpha \vDash n\}.$$

A closely related basis is the basis of fundamental quasisymmetric functions.

Definition 3.3.6. Let α be a composition. Then the *fundamental quasisymmetric function* F_α is defined by

$$F_\alpha = \sum_{\beta \preccurlyeq \alpha} M_\beta.$$

If $\alpha \vDash n$, then in terms of the variables $x_1, x_2, \dots$ we have

$$F_\alpha = \sum x_{i_1} \cdots x_{i_n},$$

where the sum is over all n-tuples $(i_1, \dots, i_n)$ of indices satisfying

$$i_1 \leqslant \cdots \leqslant i_n \text{ and } i_j < i_{j+1} \text{ if } j \in \mathrm{set}(\alpha).$$

For example, $F_{(2,1)} = M_{(2,1)} + M_{(1,1,1)}$ while $F_{(1,2)} = M_{(1,2)} + M_{(1,1,1)}$.

3.3.1 *Products and coproducts*

As with symmetric functions, the product of two quasisymmetric functions when expressed in either of the bases introduced has a combinatorial description.

Given compositions $\alpha = (\alpha_1, \dots, \alpha_k)$ and $\beta = (\beta_1, \dots, \beta_\ell)$, consider all paths P in the (x,y) plane from $(0,0)$ to (k,ℓ) with steps $(1,0)$, $(0,1)$ and $(1,1)$. Let P_i be the pointwise sum of the first i steps of P where $P_0 = (0,0)$. Then we define the composition corresponding to a path P with m steps, denoted by γ_P, to be

$$\gamma_P = (\gamma_1, \dots, \gamma_m) \tag{3.11}$$

where

$$\gamma_i = \begin{cases} \alpha_q & \text{if the } i\text{-th step is } (1,0) \text{ and } P_{i-1} = (q-1, r-1), \\ \beta_r & \text{if the } i\text{-th step is } (0,1) \text{ and } P_{i-1} = (q-1, r-1), \\ \alpha_q + \beta_r & \text{if the } i\text{-th step is } (1,1) \text{ and } P_{i-1} = (q-1, r-1). \end{cases}$$

Informally, we define the composition $\gamma_P = (\gamma_1, \dots, \gamma_m)$ corresponding to a path P with m steps as follows. Use the parts of α in order to label the unit steps on the x-axis, and use the parts of β in order to label the unit steps on the y-axis. If the i-th step of P is horizontal or vertical, then γ_i is the label of the step projected onto the x-axis or y-axis, respectively. If the i-th step of P is diagonal, then γ_i is the sum of the label of the step projected onto the x-axis plus the label of the step projected onto the y-axis. This is illustrated in the following example.

Example 3.3.7. If $\alpha = (4,5,1)$ and $\beta = (3,1)$ and P is

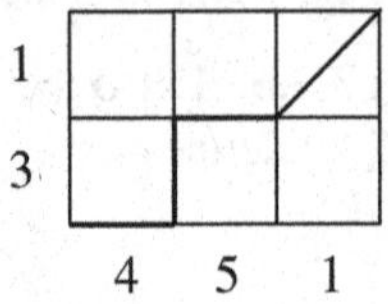

then

$$P_0 = (0,0), P_1 = (1,0), P_2 = (1,1), P_3 = (2,1), P_4 = (3,2), \text{ and } \gamma_P = (4,3,5,2).$$

The product of two monomial quasisymmetric functions is given by

$$M_\alpha M_\beta = \sum_P M_{\gamma_P} \tag{3.12}$$

where the sum is over all paths P in the (x,y) plane from $(0,0)$ to $(\ell(\alpha), \ell(\beta))$ with steps $(1,0)$, $(0,1)$ and $(1,1)$.

Given two permutations $\sigma = \sigma(1)\cdots\sigma(n) \in \mathfrak{S}_n$ and $\tau = \tau(1)\cdots\tau(m) \in \mathfrak{S}_m$, we say a *shuffle* of σ and τ is a permutation in $\mathfrak{S}_{n+m}$ such that $\sigma(i)$ appears to the right of $\sigma(i-1)$ and to the left of $\sigma(i+1)$ for all $2 \leqslant i \leqslant n-1$ and similarly, $\tau(i)+n$ appears to the right of $\tau(i-1)+n$ and to the left of $\tau(i+1)+n$ for all $2 \leqslant i \leqslant m-1$. We denote by $\sigma \sqcup\!\sqcup \tau$ the set of all shuffles of σ and τ.

Example 3.3.8.

$$12 \sqcup\!\sqcup \mathbf{21} = \{12\mathbf{43}, 1\mathbf{4}2\mathbf{3}, 1\mathbf{43}2, \mathbf{4}12\mathbf{3}, \mathbf{4}1\mathbf{3}2, \mathbf{43}12\}.$$

Let $\alpha \vDash n, \beta \vDash m$ and $\sigma \in \mathfrak{S}_n, \tau \in \mathfrak{S}_m$, such that $d(\sigma) = \mathrm{set}(\alpha)$ and $d(\tau) = \mathrm{set}(\beta)$. Then

$$F_\alpha F_\beta = \sum_{\pi \in \sigma \sqcup\!\sqcup \tau} F_{\mathrm{comp}(d(\pi))}. \tag{3.13}$$

Example 3.3.9.

$$F_{(1,2)} F_{(1)} = F_{(1,3)} + F_{(1,2,1)} + F_{(2,2)} + F_{(1,1,2)}$$

since

$$213 \in \mathfrak{S}_3 \text{ with } d(213) = \{1\} = \mathrm{set}((1,2))$$
$$1 \in \mathfrak{S}_1 \text{ with } d(1) = \emptyset = \mathrm{set}((1))$$

and

$$213 \sqcup\!\sqcup \mathbf{1} = \{213\mathbf{4}, 21\mathbf{4}3, 2\mathbf{4}13, \mathbf{4}213\}.$$

The coproduct on each of these bases is even more straightforward to describe

$$\Delta(M_\alpha) = \sum_{\alpha = \beta \cdot \gamma} M_\beta \otimes M_\gamma \qquad \Delta(F_\alpha) = \sum_{\substack{\alpha = \beta \cdot \gamma \\ \text{or } \alpha = \beta \odot \gamma}} F_\beta \otimes F_\gamma. \tag{3.14}$$

Example 3.3.10.

$$\Delta(M_{(2,1,3)}) = 1 \otimes M_{(2,1,3)} + M_{(2)} \otimes M_{(1,3)} + M_{(2,1)} \otimes M_{(3)} + M_{(2,1,3)} \otimes 1.$$

$$\begin{aligned}\Delta(F_{(2,1,3)}) = &1 \otimes F_{(2,1,3)} + F_{(1)} \otimes F_{(1,1,3)} + F_{(2)} \otimes F_{(1,3)} + F_{(2,1)} \otimes F_{(3)} \\ &+ F_{(2,1,1)} \otimes F_{(2)} + F_{(2,1,2)} \otimes F_{(1)} + F_{(2,1,3)} \otimes 1.\end{aligned}$$

The counit is given by

$$\varepsilon(M_\alpha) = \begin{cases} 1 \text{ if } \alpha = \emptyset \\ 0 \text{ otherwise} \end{cases} \tag{3.15}$$

while a formula for the antipode of the fundamental quasisymmetric function indexed by $\alpha \vDash n$ is given by

$$S(F_\alpha) = (-1)^n F_{\alpha^t}. \tag{3.16}$$

This latter formula was obtained independently by Malvenuto and Reutenauer [62, Corollary 2.3] and Ehrenborg [27, Proposition 3.4]. Furthermore, Sym is a Hopf subalgebra of QSym, and it is easily seen that for a partition λ

$$m_\lambda = \sum_{\widetilde{\alpha}=\lambda} M_\alpha. \tag{3.17}$$

For example, returning to Examples 3.2.5 and 3.3.5 we see that $m_{(2,1)} = M_{(2,1)} + M_{(1,2)}$. We can also write any skew Schur function as a linear combination of fundamental quasisymmetric functions [35, Theorem 3]

$$s_{\lambda/\mu} = \sum_\beta d_{(\lambda/\mu)\beta} F_\beta \tag{3.18}$$

where the sum is over all compositions $\beta \vDash |\lambda/\mu|$ and $d_{(\lambda/\mu)\beta}$ = the number of SYTs T of shape λ/μ such that $\mathrm{Des}(T) = \mathrm{set}(\beta)$. Equivalently,

$$s_{\lambda/\mu} = \sum_\beta d_{(\lambda/\mu)\beta} F_\beta \tag{3.19}$$

where the sum is over all compositions $\beta \vDash |\lambda/\mu|$ and $d_{(\lambda/\mu)\beta}$ = the number of SRTs $\check{T}$ of shape λ/μ such that $\mathrm{Des}(\check{T}) = \mathrm{set}(\beta)$.

Example 3.3.11. We have $s_{(3,2)} = F_{(3,2)} + F_{(2,2,1)} + F_{(2,3)} + F_{(1,3,1)} + F_{(1,2,2)}$ from the following SYTs.

4	5	
1	2	3

3	5	
1	2	4

3	4	
1	2	5

2	5	
1	3	4

2	4	
1	3	5

3.3.2 *P-partitions*

We now use the theory of P-partitions to describe the product of quasisymmetric functions. Ordinary P-partitions are a generalization of integer partitions and compositions. Our presentation is based on the summary provided in [83].

Definition 3.3.12. Let P be a finite poset. A *labelling* of P is an injective map γ from P to a chain. We call the pair (P,γ) a *labelled poset*.

Definition 3.3.13. Let (P,γ) be a labelled poset. A (P,γ)*-partition* is a map f from P to the positive integers satisfying, for all $p < q$ in P,

1. $f(p) \leqslant f(q)$, that is, f is order-preserving,
2. $f(p) = f(q)$ implies $\gamma(p) < \gamma(q)$.

We denote by $\mathscr{O}(P,\gamma)$ the set of all (P,γ)-partitions.

By a *P-partition* we shall mean a (P',γ)-partition for an arbitrary labelled poset (P',γ).

We note that the conditions of Definition 3.3.13 are satisfied for all $p < q$ as soon as they are satisfied for all coverings $p \lessdot q$, since $p < q$ in the finite poset P implies that there are elements $p_1, \ldots, p_k \in P$ such that

$$p \lessdot p_1 \lessdot \cdots \lessdot p_k \lessdot q.$$

Then the first condition implies

$$f(p) \leqslant f(p_1) \leqslant \cdots \leqslant f(p_k) \leqslant f(q).$$

In particular, if $f(p) = f(q)$, then

$$f(p) = f(p_1) = \cdots = f(p_k) = f(q),$$

hence the second condition implies

$$\gamma(p) < \gamma(p_1) < \cdots < \gamma(p_k) < \gamma(q).$$

Stanley's definition of P-partitions in [77] differs from the one given here by requiring them to be order-reversing rather than order-preserving, but the theories obtained from the two definitions are equivalent: To obtain one from the other, one need only replace a poset P by its dual P^*.

We shall give an example of a (P, γ)-partition where P is a chain, since these are the posets that will most interest us. In our example, the labelled chain will be represented by the following type of diagram.

Definition 3.3.14. Let (w, γ) be a labelled chain with order $w_1 < w_2 < \cdots$. The *zigzag diagram* of (w, γ) is the graph with vertices w_i and edges (w_i, w_{i+1}), drawn so that the vertex w_{i+1} is

1. to the right of and above the vertex w_i if $\gamma(w_i) < \gamma(w_{i+1})$,
2. to the right of and below the vertex w_i if $\gamma(w_i) > \gamma(w_{i+1})$.

Example 3.3.15. Let (w, γ) be the labelled chain with order $w_1 < \cdots < w_4$ and labelling γ that maps $w_1 \mapsto 5$, $w_2 \mapsto 2$, $w_3 \mapsto 7$ and $w_4 \mapsto 8$. Below each vertex w_i of the zigzag diagram of (w, γ), we have written the value $f(w_i)$ of a (w, γ)-partition f.

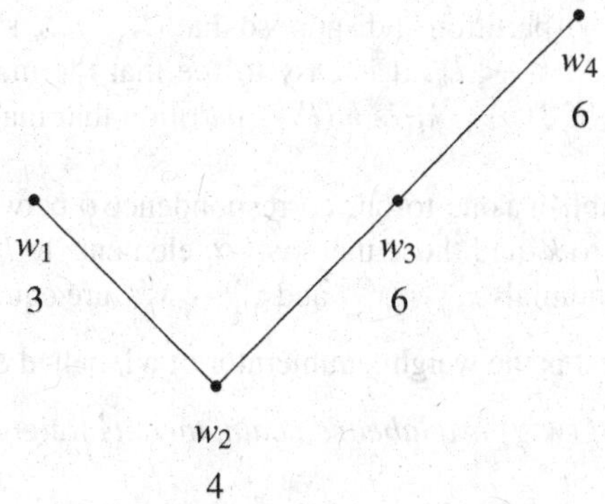

The appearance of a zigzag diagram suggests the following terminology.

Definition 3.3.16. Let (w, γ) be a labelled chain with order $w_1 < w_2 < \cdots$. We say the number i is a *descent* of (w, γ) if $\gamma(w_i) > \gamma(w_{i+1})$, and an *ascent* if $\gamma(w_i) < \gamma(w_{i+1})$.

The set of all descents of (w, γ) is denoted by $D(w, \gamma)$.

Example 3.3.17. The labelled chain (w, γ) of Example 3.3.15 has descent 1 and ascents 2 and 3.

Using the terminology just introduced and taking into account the remark about coverings in the paragraph following Definition 3.3.13, we can describe a (w, γ)-partition as a map f from w to the positive integers satisfying

1. $f(w_i) \leqslant f(w_{i+1})$,
2. $f(w_i) = f(w_{i+1})$ implies i is an ascent of (w, γ) or equivalently, i is a descent of (w, γ) implies $f(w_i) < f(w_{i+1})$.

We now introduce generating functions for P-partitions.

Definition 3.3.18. Let (P, γ) be a labelled poset. For any (P, γ)-partition f, denote by x^f the monomial

$$x^f = \prod_{p \in P} x_{f(p)}.$$

Then the *weight enumerator* of (P, γ) is the formal power series $F(P, \gamma)$ defined by

$$F(P, \gamma) = \sum x^f,$$

where the sum is over all (P, γ)-partitions f.

The generating functions just introduced are quasisymmetric.

Proposition 3.3.19. *The weight enumerator $F(P, \gamma)$ of a labelled poset (P, γ) is a quasisymmetric function.*

Proof. Let $(\alpha_1, \dots, \alpha_m)$ be a composition of $|P|$ and $(k_1, \dots, k_m)$ a sequence of positive integers $k_1 < \cdots < k_m$. The coefficient of the monomial $x_{k_1}^{\alpha_1} \cdots x_{k_m}^{\alpha_m}$ in $F(P, \gamma)$ is the number of (P, γ)-partitions that map α_i elements of P to k_i.

Let f be such a (P, γ)-partition and suppose that $(l_1, \dots, l_m)$ is another sequence of positive integers $l_1 < \cdots < l_m$. It is easy to see that the map $\phi(f)$, defined by setting $(\phi(f))(p) = l_i$ if $f(p) = k_i$, is a (P, γ)-partition that maps α_i elements of P to l_i.

In this way, we establish a one-to-one correspondence ϕ between (P, γ)-partitions that map α_i elements to k_i and those that map α_i elements to l_i. It follows that the coefficients of the monomials $x_{k_1}^{\alpha_1} \cdots x_{k_m}^{\alpha_m}$ and $x_{l_1}^{\alpha_1} \cdots x_{l_m}^{\alpha_m}$ are equal. □

Of particular interest is the weight enumerator of a labelled chain.

Proposition 3.3.20. *If (w, γ) is a labelled chain, then its weight enumerator*

$$F(w,\gamma) = F_\alpha,$$

the fundamental quasisymmetric function indexed by the composition $\alpha \vDash |w|$ *with* $\mathrm{set}(\alpha) = D(w,\gamma)$.

Proof. By the paragraph after Definition 3.3.6 we have

$$F_\alpha = \sum x_{i_1} \cdots x_{i_n},$$

where the sum is over all sequences $(i_1,\ldots,i_n)$ of positive integers satisfying

$$i_1 \leqslant \cdots \leqslant i_n \text{ and } i_j < i_{j+1} \text{ if } j \in \mathrm{set}(\alpha) = D(w,\gamma).$$

Now

$$F(w,\gamma) = \sum_{f\in\mathcal{O}(w,\gamma)} x_{f(w_1)} \cdots x_{f(w_n)}.$$

By the remarks in the paragraph following Example 3.3.17, we have $f \in \mathcal{O}(w,\gamma)$ if and only if

$$f(w_1) \leqslant \cdots \leqslant f(w_n) \text{ and } f(w_j) < f(w_{j+1}) \text{ if } j \in D(w,\gamma).$$

Since the values in the range of a (w,γ)-partition can be any positive integers, the result follows. □

We have just shown that every weight enumerator of a labelled chain is a fundamental quasisymmetric function. The converse is also true: Given a fundamental quasisymmetric function F_α, we can always find a labelled chain (w,γ) such that $F_\alpha = F(w,\gamma)$.

Indeed, let w be the chain with order $w_1 < \cdots < w_n$, where $n = |\alpha|$. If $i_1 < i_2 < \cdots$ are the elements of $\mathrm{set}(\alpha)$ and $j_1 < j_2 < \cdots$ are the elements of $[n] - \mathrm{set}(\alpha)$, let γ respectively map

$$w_{i_1}, w_{i_2}, \ldots \mapsto n, n-1, \ldots \text{ and } w_{j_1}, w_{j_2}, \ldots \mapsto 1,2,\ldots.$$

Then $D(w,\gamma) = \mathrm{set}(\alpha)$, since

1. $i, i+1 \in \mathrm{set}(\alpha)$ implies $\gamma(w_i) > \gamma(w_{i+1})$,
2. $j, j+1 \in [n] - \mathrm{set}(\alpha)$ implies $\gamma(w_j) < \gamma(w_{j+1})$,
3. $i \in \mathrm{set}(\alpha)$ and $j \in [n] - \mathrm{set}(\alpha)$ implies $\gamma(w_i) > \gamma(w_j)$.

Example 3.3.21. If α is the composition $(3,2,4) \vDash 9$, then

$$\mathrm{set}(\alpha) = \{3,5\} \text{ and } [n] - \mathrm{set}(\alpha) = \{1,2,4,6,7,8,9\}.$$

Let w be the chain with order $w_1 < \cdots < w_9$ and labelling γ that respectively maps

$$w_3, w_5 \mapsto 9,8 \text{ and } w_1, w_2, w_4, w_6, w_7, w_8, w_9 \mapsto 1,2,3,4,5,6,7.$$

Then γ respectively maps

$$w_1,\ldots,w_9 \mapsto 1,2,9,3,8,4,5,6,7,$$

hence $D(w,\gamma)=\{3,5\}$.

Thus the fundamental basis of QSym consists of generating functions for P-partitions. Observe that the Schur basis of Sym can be viewed as generating functions for semistandard Young tableaux. In fact, P-partitions are a generalization of SSYTs: An SSYT of shape λ is just a P-partition of the labelled poset $(P_\lambda,\gamma_\lambda)$ defined as follows.

Let P_λ be the poset whose elements are the pairs (i,j) of row and column coordinates for cells in the Young diagram of λ, with order defined by

$$(i,j)\leqslant(k,l) \text{ if } i\leqslant k \text{ and } j\leqslant l.$$

In particular, $(i,j)<(i,j+1)$ and $(i,j)<(i+1,j)$, so elements of P_λ increase in order as we move from left to right in rows and bottom to top in columns.

Let γ_λ be the labelling that assigns the numbers $1,2,\ldots$ to the elements of P_λ in the following order: first the coordinates of the cells in the first column, from top to bottom; then the coordinates of the cells in the second column, from top to bottom; and so on.

Example 3.3.22. Let $\lambda=(5,4,2,2)$. Below is the Young diagram of λ, with the label $\gamma_\lambda(i,j)$ written in the cell with coordinate pair (i,j).

1	5			
2	6			
3	7	9	11	
4	8	10	12	13

Let T be an SSYT of shape λ. If we regard T as the map from P_λ to the positive integers that sends the pair (i,j) to the entry of the cell with coordinate pair (i,j), then T is a $(P_\lambda,\gamma_\lambda)$-partition. Moreover, by Definition 3.2.8 and Definition 3.3.18 it immediately follows that

$$F(P_\lambda,\gamma_\lambda)=s_\lambda. \tag{3.20}$$

We now resume our development of the theory of P-partitions. The next lemma will allow us to write any weight enumerator as a sum of weight enumerators of chains.

Lemma 3.3.23 (Fundamental lemma of P-Partitions). *If (P,γ) is a labelled poset, then the set of all (P,γ)-partitions*

$$\mathscr{O}(P,\gamma)=\dot{\bigcup}\mathscr{O}(w,\gamma),$$

where the disjoint union is over all linear extensions w of P.

Proof. First note that a labelling of P is also a labelling of any linear extension of P.

Now suppose that f is a (P,γ)-partition. Let w be the chain with underlying set P and order defined by $p < q$ in w if

1. $f(p) < f(q)$ or
2. $f(p) = f(q)$ and $\gamma(p) < \gamma(q)$.

If $p < q$ in P, then Definition 3.3.13 ensures that one of the two conditions is satisfied. Thus w is a linear extension of P, and it is clearly the only one for which f is a (w,γ)-partition.

The converse is trivial: If w is a linear extension of P, then every (w,γ)-partition is also a (P,γ)-partition. □

The following result is an immediate consequence of Lemma 3.3.23.

Corollary 3.3.24. *If (P,γ) is a labelled poset, then its weight enumerator*

$$F(P,\gamma) = \sum F(w,\gamma),$$

where the sum is over all linear extensions w of P.

As an application, we will now use Corollary 3.3.24 and Equation (3.20) to deduce Equation (3.18) in the case $\mu = \emptyset$. First note that if we return to our poset P_λ then a linear extension w of P_λ corresponds to an SYT T_w with $sh(T_w) = \lambda$. More precisely, $T_w(i,j)$ is $1+$ rank of (i,j) in w. By Proposition 3.3.20 $F(w,\gamma_\lambda) = F_\alpha$ where $\alpha \vDash |w|$ and $\mathrm{set}(\alpha) = D(w,\gamma_\lambda)$. However, $D(w,\gamma_\lambda)$ consists precisely of all i where $i+1$ appears weakly left of i in T_w, that is, $D(w,\gamma_\lambda) = \mathrm{Des}(T_w)$. Therefore by Corollary 3.3.24 and Equation (3.20)

$$s_\lambda = \sum F_{\mathrm{comp}(\mathrm{Des}(T))}$$

where the sum is over all SYTs T with $sh(T) = \lambda$.

We can also use the theory of P-partitions to derive a multiplication rule for quasisymmetric functions. Given fundamental quasisymmetric functions F_α and F_β, choose labelled chains (u,γ) and (v,δ) such that

1. u and v are disjoint sets,
2. $|u| = |\alpha|$ and $|v| = |\beta|$,
3. $\gamma(u)$ and $\delta(v)$ are disjoint subsets of the same chain,
4. $D(u,\gamma) = \mathrm{set}(\alpha)$ and $D(v,\delta) = \mathrm{set}(\beta)$.

We can construct such labelled chains using the procedure described in the paragraph preceding Example 3.3.21, then adding $|\alpha|$ to each label of the second chain.

From Definition 3.3.18, it is clear that

$$F(u,\gamma)F(v,\delta) = F(u+v,\gamma+\delta),$$

where $\gamma+\delta$ is the labelling of the disjoint union $u+v$ that maps $u_i \mapsto \gamma(u_i)$ and $v_j \mapsto \delta(v_j)$. Thus

$$\begin{aligned} F_\alpha F_\beta &= F(u,\gamma)F(v,\delta) \\ &= F(u+v,\gamma+\delta) \\ &= \sum_{w\in\mathscr{L}(u+v)} F(w,\gamma+\delta) \\ &= \sum_{w\in\mathscr{L}(u+v)} F_{\alpha(w)}, \end{aligned}$$

where $\alpha(w) \vDash |w|$ satisfies $\mathrm{set}(\alpha(w)) = D(w,\gamma+\delta)$. Observe this is equivalent to Equation (3.13).

We have just seen that the product of quasisymmetric functions corresponds to operations on labelled posets. We shall now see that the same is true for coproduct and the antipode.

From (3.14) we obtain the formula

$$\Delta(F_\alpha) = F_\alpha \otimes 1 + \sum F_{\zeta\cdot(a)} \otimes F_{(b)\cdot\eta},$$

where the sum is over all ways of writing $\alpha = \zeta\cdot(a+b)\cdot\eta$, a concatenation of compositions where $a \geqslant 0$ and $b > 0$ are integers adding up to a part of α, and we set $(a) = \emptyset$ if $a = 0$.

Let (w,γ) be a labelled chain with $|w| = |\alpha|$ and descent set $D(w,\gamma) = \mathrm{set}(\alpha)$. Then every tensor $F_\mu \otimes F_\nu$ appearing in the expansion of the coproduct $\Delta(F_\alpha)$ corresponds to cutting the labelled chain (w,γ) into two labelled chains with descent sets respectively $\mathrm{set}(\mu)$ and $\mathrm{set}(\nu)$.

Example 3.3.25. We have

$$\begin{aligned} \Delta(F_{(3,2,4)}) = F_{(3,2,4)} \otimes 1 &+ F_{(3,2,3)} \otimes F_{(1)} + F_{(3,2,2)} \otimes F_{(2)} + F_{(3,2,1)} \otimes F_{(3)} \\ &+ F_{(3,2)} \otimes F_{(4)} + F_{(3,1)} \otimes F_{(1,4)} + F_{(3)} \otimes F_{(2,4)} + F_{(2)} \otimes F_{(1,2,4)} \\ &+ F_{(1)} \otimes F_{(2,2,4)} + 1 \otimes F_{(3,2,4)}. \end{aligned}$$

A labelled chain (w,γ) with

$$|w| = |(3,2,4)| = 9 \text{ and } D(w,\gamma) = \mathrm{set}((3,2,4)) = \{3,5\}$$

has the following zigzag diagram.

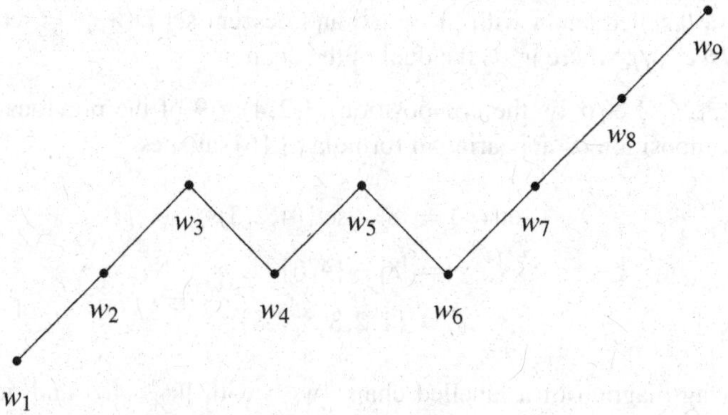

Consider the tensor $F_{(3,1)} \otimes F_{(1,4)}$ in the expansion of $\Delta(F_{(3,2,4)})$. Below are the zigzag diagrams of a labelled chain (u,δ) with

$$|u| = |(3,1)| = 4 \text{ and } D(u,\delta) = \mathrm{set}((3,1)) = \{3\},$$

and a labelled chain (v,ζ) with

$$|v| = |(1,4)| = 5 \text{ and } D(v,\zeta) = \mathrm{set}((1,4)) = \{1\}.$$

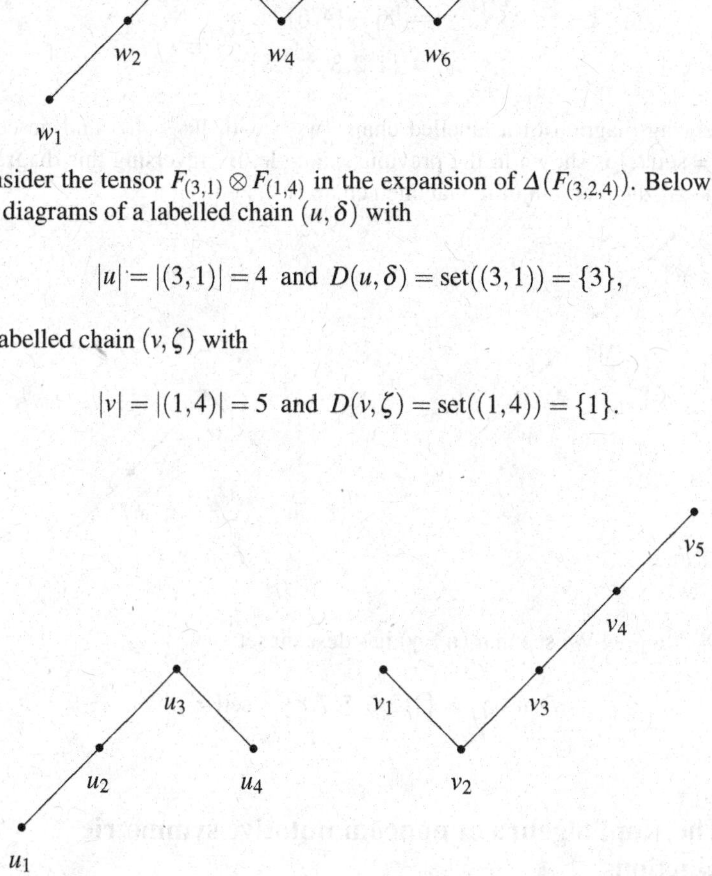

As for the antipode, recall that formula (3.16) gives the antipode of the fundamental quasisymmetric function indexed by $\alpha = (\alpha_1, \ldots, \alpha_k) \vDash n$ as

$$S(F_\alpha) = (-1)^n F_{\alpha^t},$$

where we recall $\alpha^t \vDash n$ satisfies

$$\mathrm{set}(\alpha^t) = [n-1] - \mathrm{set}(\alpha_k, \ldots, \alpha_1).$$

If (w,γ) is a labelled chain with $|w| = |\alpha|$ and descent set $D(w,\gamma) = \mathrm{set}(\alpha)$, then $\mathrm{set}(\alpha^t) = D(w^*,\gamma)$, where w^* is the dual of the chain w.

Example 3.3.26. Let α be the composition $(3,2,4) \vDash 9$ of the previous example. Then the composition α^t appearing in formula (3.16) satisfies

$$\begin{aligned}\mathrm{set}(\alpha^t) &= [8] - \mathrm{set}((4,2,3)) \\ &= [8] - \{4,6\} \\ &= \{1,2,3,5,7,8\}.\end{aligned}$$

The zigzag diagram of a labelled chain (w,γ) with $|w| = |\alpha|$ and descent set $D(w,\gamma) = \mathrm{set}(\alpha)$ is shown in the previous example. By reversing this diagram we obtain the zigzag diagram of the labelled chain (w^*,γ)

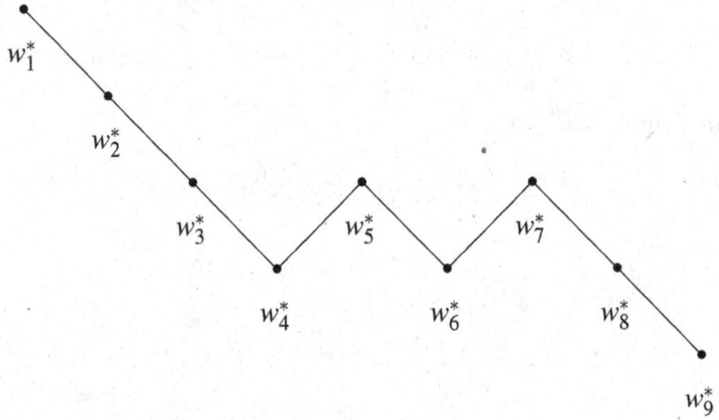

where $w_i^* = w_{10-i}$. We see that (w^*,γ) has descent set

$$D(w^*,\gamma) = \{1,2,3,5,7,8\} = \mathrm{set}(\alpha^t).$$

3.4 The Hopf algebra of noncommutative symmetric functions

Our third and final Hopf algebra involves noncommutative symmetric functions, defined by Gelfand et al. in [34]. As we will see, they are closely connected to both quasisymmetric and symmetric functions. Throughout we use the notation used in [11], which evokes the relationship with symmetric functions.

Definition 3.4.1. The Hopf algebra of *noncommutative symmetric functions*, denoted by NSym, is

$$\mathbb{Q}\langle \mathbf{e}_1, \mathbf{e}_2, \ldots \rangle$$

generated by noncommuting indeterminates $\mathbf{e}_n$ of degree n with the operations of the next subsection.

This definition is analogous to the earlier fundamental theorem of symmetric functions, which showed that we can regard Sym as the algebra

$$\mathbb{Q}[e_1, e_2, \ldots]$$

generated by commuting indeterminates e_n of degree n. The set of all noncommutative symmetric functions forms a graded Hopf algebra

$$\mathrm{NSym} = \bigoplus_{n \geqslant 0} \mathrm{NSym}^n$$

where NSym is spanned by the following functions $\mathbf{e}_\alpha$.

Definition 3.4.2. Let n be a nonnegative integer. Then the *n-th elementary noncommutative symmetric function*, denoted by $\mathbf{e}_n$, is the indeterminate $\mathbf{e}_n$ where we set $\mathbf{e}_0 = 1$. The *n-th complete homogeneous noncommutative symmetric function*, denoted by $\mathbf{h}_n$, is defined by $\mathbf{h}_0 = 1$

$$\mathbf{h}_n = \sum_{(y_1, y_2, \ldots, y_m) \vDash n} (-1)^{n-m} \mathbf{e}_{y_1} \mathbf{e}_{y_2} \cdots \mathbf{e}_{y_m}.$$

Let $\alpha = (\alpha_1, \ldots, \alpha_k)$ be a composition. The *elementary noncommutative symmetric function* $\mathbf{e}_\alpha$ is defined by

$$\mathbf{e}_\alpha = \mathbf{e}_{\alpha_1} \cdots \mathbf{e}_{\alpha_k}$$

and the *complete homogeneous noncommutative symmetric function* $\mathbf{h}_\alpha$ is defined by

$$\mathbf{h}_\alpha = \mathbf{h}_{\alpha_1} \cdots \mathbf{h}_{\alpha_k} = \sum_{\beta \preccurlyeq \alpha} (-1)^{|\alpha| - \ell(\beta)} \mathbf{e}_\beta.$$

The *noncommutative ribbon Schur function* is defined by

$$\mathbf{r}_\alpha = \sum_{\beta \succcurlyeq \alpha} (-1)^{\ell(\alpha) - \ell(\beta)} \mathbf{h}_\beta.$$

Example 3.4.3. We have $\mathbf{h}_{(2,1)} = -\mathbf{e}_{(2,1)} + \mathbf{e}_{(1,1,1)}$ and $\mathbf{h}_{(1,2)} = -\mathbf{e}_{(1,2)} + \mathbf{e}_{(1,1,1)}$ while $\mathbf{r}_{(1,2,1)} = \mathbf{h}_{(1,2,1)} - \mathbf{h}_{(3,1)} - \mathbf{h}_{(1,3)} + \mathbf{h}_{(4)}$.

The $\mathbf{e}_\alpha$ are clearly independent, and the matrix expressing the $\mathbf{h}_\alpha$ in terms of the $\mathbf{e}_\beta$ and the matrix expressing the $\mathbf{r}_\alpha$ in terms of the $\mathbf{h}_\beta$ are upper triangular with ± 1 on the diagonal. Therefore if we index the rows and columns of these matrices using a linear extension of refinement we can deduce

$$\mathrm{NSym}^n = \mathrm{span}\{\mathbf{e}_\alpha \mid \alpha \vDash n\} = \mathrm{span}\{\mathbf{h}_\alpha \mid \alpha \vDash n\} = \mathrm{span}\{\mathbf{r}_\alpha \mid \alpha \vDash n\}.$$

3.4.1 Products and coproducts

With each of the bases introduced the product of two such functions is not hard to describe. For compositions α and β we have

$$\mathbf{e}_\alpha \mathbf{e}_\beta = \mathbf{e}_{\alpha\cdot\beta} \tag{3.21}$$

$$\mathbf{h}_\alpha \mathbf{h}_\beta = \mathbf{h}_{\alpha\cdot\beta} \tag{3.22}$$

and

$$\mathbf{r}_\alpha \mathbf{r}_\beta = \mathbf{r}_{\alpha\cdot\beta} + \mathbf{r}_{\alpha\odot\beta}. \tag{3.23}$$

For example, $\mathbf{h}_{(1,4,1,2)}\mathbf{h}_{(3,1,1)} = \mathbf{h}_{(1,4,1,2,3,1,1)}$ and $\mathbf{r}_{(1,4,1,2)}\mathbf{r}_{(3,1,1)} = \mathbf{r}_{(1,4,1,2,3,1,1)} + \mathbf{r}_{(1,4,1,5,1,1)}$.

Meanwhile, the coproduct is easy to describe for the elementary and complete homogeneous noncommutative symmetric functions

$$\Delta(\mathbf{e}_n) = \sum_{i=0}^{n} \mathbf{e}_i \otimes \mathbf{e}_{n-i} \quad \Delta(\mathbf{h}_n) = \sum_{i=0}^{n} \mathbf{h}_i \otimes \mathbf{h}_{n-i}.$$

The counit is given by

$$\varepsilon(\mathbf{e}_\alpha) = \begin{cases} 1 \text{ if } \alpha = \emptyset \\ 0 \text{ otherwise} \end{cases} \tag{3.24}$$

while a formula for the antipode of the n-th complete homogeneous noncommutative symmetric function is given by

$$S(\mathbf{h}_n) = (-1)^n \mathbf{e}_n. \tag{3.25}$$

3.4.2 Duality

The work of [34] combined with that of Gessel [35] and Malvenuto and Reutenauer [62] showed that NSym is the graded Hopf dual of QSym. In [34] a pairing was introduced that pairs the elements of any basis of QSym with its dual basis in NSym and satisfies

$$\langle M_\alpha, \mathbf{h}_\beta \rangle = \delta_{\alpha\beta} \tag{3.26}$$

and

$$\langle F_\alpha, \mathbf{r}_\beta \rangle = \delta_{\alpha\beta}. \tag{3.27}$$

By duality, the inclusion Sym $\hookrightarrow$ QSym induces a quotient map

$$\chi : \mathrm{NSym} \to \mathrm{Sym} \tag{3.28}$$

satisfying $\chi(\mathbf{e}_n) = e_n$. The map χ is often called the *forgetful map* and can be thought of as allowing the natural indeterminates in the image to commute. Under the forgetful map we have for a composition α that

$$\chi(\mathbf{h}_\alpha) = h_{\tilde{\alpha}} \tag{3.29}$$

$$\chi(\mathbf{r}_\alpha) = r_\alpha. \tag{3.30}$$

3.5 Relationship between Sym, QSym, and NSym

We summarize the relationship between Sym, QSym, and NSym with the following diagram.

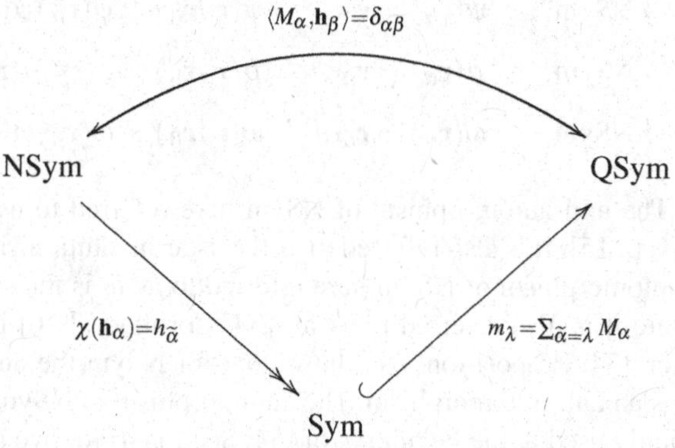

3.6 Automorphisms

The notions of complement, reversal, and transpose of compositions correspond to well-known involutive automorphisms of QSym, which can be defined in terms of the fundamental basis as follows.

$$\psi : \mathrm{QSym} \to \mathrm{QSym}, \qquad \psi(F_\alpha) = F_{\alpha^c} \tag{3.31}$$

$$\rho : \mathrm{QSym} \to \mathrm{QSym}, \qquad \rho(F_\alpha) = F_{\alpha^r} \tag{3.32}$$

$$\omega : \mathrm{QSym} \to \mathrm{QSym}, \qquad \omega(F_\alpha) = F_{\alpha^t} \tag{3.33}$$

Note that these automorphisms commute, and that $\omega = \rho \circ \psi = \psi \circ \rho$. Moreover, considering QSym as a Hopf algebra, the antipode S is also given by $S(F_\alpha) = (-1)^{|\alpha|}\omega(F_\alpha)$. The automorphism ρ restricts to the identity on Sym, and both ω

and ψ restrict to the well-known "conjugating" automorphism of Sym, which is usually denoted by ω and satisfies $\omega(h_r) = e_r$.

Remark 3.6.1. Ehrenborg in [27, p. 6] refers to the automorphism we here refer to as ρ, there using the notation $f \mapsto f^*$. The automorphism of QSym here referred to as ω is the same as that also referred to as ω in the papers of Malvenuto and Reutenauer [62, p. 975] and [63, Section 3]. Ehrenborg [27, Section 5] also defines an involution of QSym which he calls ω, but Ehrenborg's ω is what we call ψ. This can be confusing since the papers [27] and [62] refer to each other regarding "ω" but do not mention this distinction. Stanley in [81, Exercise 7.94] also refers to the involution ψ, there referred to as $\hat{\omega}$.

There are corresponding involutions of NSym, defined in terms of the noncommutative ribbon Schur basis [34]. However, since NSym is noncommutative, we take care to note that ψ is an automorphism whereas ρ and ω are anti-automorphisms.

$$\psi : \mathrm{NSym} \to \mathrm{NSym}, \qquad \psi(\mathbf{r}_\alpha) = \mathbf{r}_{\alpha^c}, \qquad \psi(\mathbf{r}_\alpha \mathbf{r}_\beta) = \psi(\mathbf{r}_\alpha)\psi(\mathbf{r}_\beta) \tag{3.34}$$

$$\rho : \mathrm{NSym} \to \mathrm{NSym}, \qquad \rho(\mathbf{r}_\alpha) = \mathbf{r}_{\alpha^r}, \qquad \rho(\mathbf{r}_\alpha \mathbf{r}_\beta) = \rho(\mathbf{r}_\beta)\rho(\mathbf{r}_\alpha) \tag{3.35}$$

$$\omega : \mathrm{NSym} \to \mathrm{NSym}, \qquad \omega(\mathbf{r}_\alpha) = \mathbf{r}_{\alpha^t}, \qquad \omega(\mathbf{r}_\alpha \mathbf{r}_\beta) = \omega(\mathbf{r}_\beta)\omega(\mathbf{r}_\alpha) \tag{3.36}$$

Remark 3.6.2. The anti-automorphism of NSym here referred to as ρ is denoted by$f \mapsto f^*$ in [34, p. 15]. It is also referred to as the "star involution" in [51, Section 2.3]. The anti-automorphism of NSym here referred to as ω is the same as that in [34, p. 18], where it is also referred to as ω; [34, Corollary 3.16] is our formula (3.36). Moreover, [34, Proposition 3.9] shows that for NSym the antipode, which they denote $\tilde{\omega}$, is an anti-automorphism. The automorphism of NSym here referred to as ψ is implicitly used at the beginning of [34, Section 4] to give an abbreviated description of some of the transition matrices between bases of NSym, for example, $\psi(\mathbf{h}_\alpha) = \mathbf{e}_\alpha$.

For each of these involutions it is worth noting what are the images of bases of interest. For example, on Sym we have $\omega(h_\lambda) = e_\lambda$ and $\omega(m_\lambda) = f_\lambda$, where $\{f_\lambda\}$ is the basis of *forgotten symmetric functions*. On NSym, we have $\rho(\mathbf{h}_\alpha) = \mathbf{h}_{\alpha^r}$, $\omega(\mathbf{h}_\alpha) = \mathbf{e}_{\alpha^r}$ [34, Equation (44)], and $\psi(\mathbf{h}_\alpha) = \mathbf{e}_\alpha$. Similarly, on QSym, $\rho(M_\alpha) = M_{\alpha^r}$. On QSym, the image of the monomial basis under ω does not have a standard name, but would be the dual of the $\{\mathbf{e}_\alpha\}$ basis of NSym.

In summary, under all these involutions, the fundamental basis of QSym is preserved (though permuted), in the sense that, say, $\{\omega(F_\alpha)\}_\alpha = \{F_\alpha\}_\alpha$, and likewise the noncommutative ribbon Schur basis is preserved. We see that the monomial basis of QSym and the complete homogeneous noncommutative basis of NSym are preserved under ρ, but not under the other involutions. For example,

$$\rho(\mathbf{h}_{21}) = \rho(\mathbf{h}_2\mathbf{h}_1) = \rho(\mathbf{h}_1)\rho(\mathbf{h}_2) = \mathbf{h}_1\mathbf{h}_2 = \mathbf{h}_{12}$$
$$\omega(\mathbf{h}_{21}) = \omega(\mathbf{h}_2\mathbf{h}_1) = \omega(\mathbf{h}_1)\omega(\mathbf{h}_2) = \mathbf{e}_1\mathbf{e}_2 = \mathbf{e}_{12} = \mathbf{h}_{111} - \mathbf{h}_{12}$$
$$\psi(\mathbf{h}_{21}) = \psi(\mathbf{h}_2\mathbf{h}_1) = \psi(\mathbf{h}_2)\psi(\mathbf{h}_1) = \mathbf{e}_2\mathbf{e}_1 = \mathbf{e}_{21} = \mathbf{h}_{111} - \mathbf{h}_{21}.$$

3.7 Combinatorial Hopf algebras

The idea that certain Hopf algebras provide a natural setting for the study of combinatorial problems was formalized by Aguiar, Bergeron and Sottile [4]. They defined a *combinatorial Hopf algebra* (abbreviated to *CH-algebra*) to be a pair $(\mathscr{H}, \zeta)$ consisting of a connected, graded Hopf algebra

$$\mathscr{H} = \bigoplus_{n\geqslant 0} \mathscr{H}_n$$

over a field K with finite-dimensional components $\mathscr{H}_n$, and a multiplicative functional $\zeta : \mathscr{H} \to K$. The functional ζ may be considered as a generalization of the classical zeta function defined on the intervals of a poset, which can be used to count chains in the poset.

A *CH-morphism* between CH-algebras $(\mathscr{H}, \zeta)$ and $(\mathscr{H}', \zeta')$ is a Hopf morphism $\psi : \mathscr{H} \to \mathscr{H}'$ such that $\psi(\mathscr{H}_n) \subseteq \mathscr{H}'_n$ for all n, and the following diagram commutes.

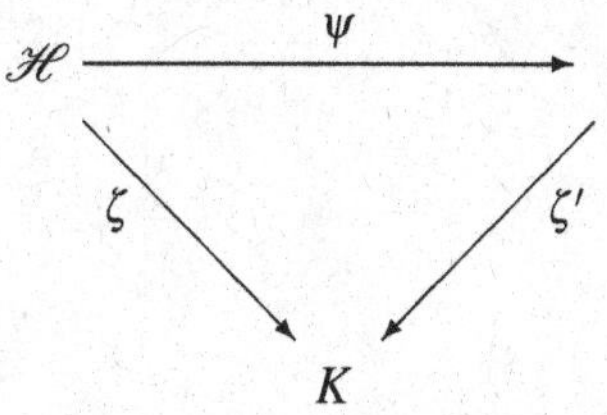

The algebra QSym plays a special role in the theory of CH-algebras. By allowing the coefficients of quasisymmetric functions to be elements of the field K, we obtain the Hopf algebra $K \otimes \mathrm{QSym}$ over K. A multiplicative functional $\eta : K \otimes \mathrm{QSym} \to K$ is given by

$$\eta(M_\alpha) = \begin{cases} 1 \text{ if } \alpha = \emptyset \text{ or } (n) \\ 0 \text{ otherwise.} \end{cases}$$

Aguiar, Bergeron and Sottile adapted a result by Aguiar [1] concerning infinitesimal Hopf algebras to prove the following.

Theorem 3.7.1. *[4] Let* $\eta : K \otimes \mathrm{QSym} \to K$ *be the functional given in the previous paragraph. Then for any CH-algebra* $(\mathscr{H}, \zeta)$*, there exists a unique CH-morphism* $\psi : \mathscr{H} \to K \otimes \mathrm{QSym}$*, that is, a Hopf morphism* ψ *such that* $\psi(\mathscr{H}_n) \subseteq (K \otimes \mathrm{QSym})_n$ *for all n, and the following diagram commutes.*

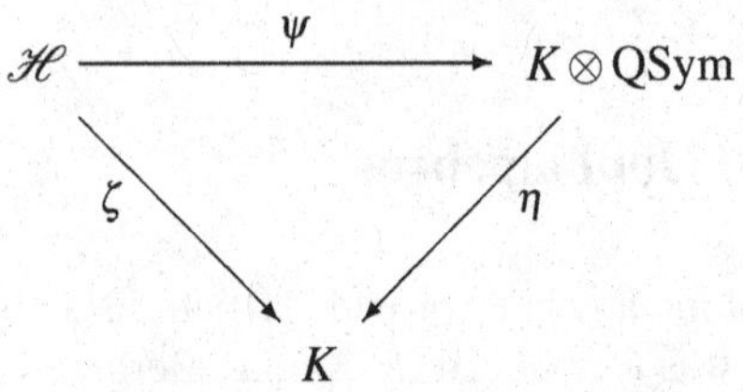

In the language of category theory, this theorem asserts that the CH-algebra $(K \otimes \mathrm{QSym}, \eta)$ is the terminal object in the category of CH-algebras over K and their CH-morphisms.

Chapter 4
Composition tableaux and further combinatorial concepts

Abstract In order to state results in the next chapter, we extend many definitions from Chapter 2 to define composition diagrams, Young composition tableaux that correspond to Young tableaux, and the Young composition poset. We additionally define reverse composition diagrams, reverse composition tableaux that correspond to reverse tableaux, and the reverse composition poset. Finally, useful bijections between Young tableaux, Young composition tableaux, reverse tableaux and reverse composition tableaux are described.

4.1 Young composition tableaux and the Young composition poset

It is important to note that the combinatorial concepts introduced in Section 4.2 are those used in [15, 40]. However the analogous combinatorial concepts introduced here are related to Young tableaux and hence will enable stronger parallels to be drawn with classical results in the final chapter.

Definition 4.1.1. Given a composition $\alpha = (\alpha_1, \dots, \alpha_{\ell(\alpha)}) \vDash n$, we say the *Young composition diagram* of α, also denoted by α, is the left-justified array of n cells with α_i cells in the i-th row. We follow the Cartesian or French convention, which means that we number the rows from bottom to top, and the columns from left to right. The cell in the i-th row and j-th column is denoted by the pair (i, j).

K. Luoto et al., *An Introduction to Quasisymmetric Schur Functions*,
SpringerBriefs in Mathematics, DOI 10.1007/978-1-4614-7300-8_4,
© Kurt Luoto, Stefan Mykytiuk, Stephanie van Willigenburg 2013

Example 4.1.2.

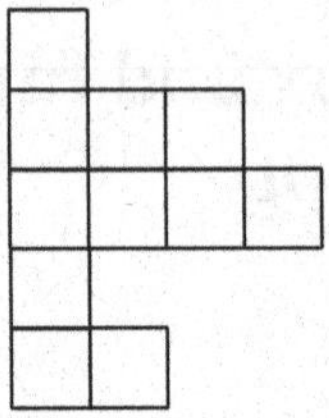

$$\alpha = (2,1,4,3,1)$$

We now define a poset to enable us to define skew versions of Young composition diagrams.

Definition 4.1.3. The *Young composition poset* $\mathscr{L}_{\hat{c}}$ is the poset consisting of all compositions in which $\alpha = (\alpha_1, \ldots, \alpha_\ell)$ is covered by

1. $(\alpha_1, \ldots, \alpha_\ell, 1)$, that is, the composition obtained by *suffixing* a part of size 1 to α.
2. $(\alpha_1, \ldots, \alpha_k + 1, \ldots, \alpha_\ell)$, provided that $\alpha_i \neq \alpha_k$ for all $i > k$, that is, the composition obtained by adding 1 to a part of α as long as that part is the *rightmost* part of that size.

As with Young's lattice, we can note the *column sequence* of a saturated chain in $\mathscr{L}_{\hat{c}}$.

Example 4.1.4. A saturated chain in $\mathscr{L}_{\hat{c}}$ is

$$(1) \lessdot_{\hat{c}} (1,1) \lessdot_{\hat{c}} (1,2) \lessdot_{\hat{c}} (2,2) \lessdot_{\hat{c}} (2,3) \lessdot_{\hat{c}} (2,3,1)$$

and

$$\mathrm{col}(\,(1) \lessdot_{\hat{c}} (1,1) \lessdot_{\hat{c}} (1,2) \lessdot_{\hat{c}} (2,2) \lessdot_{\hat{c}} (2,3) \lessdot_{\hat{c}} (2,3,1)\,) \;=\; 1,2,2,3,1.$$

Let α, β be two Young composition diagrams such that $\beta <_{\hat{c}} \alpha$. Then we define the *skew Young composition shape* $\alpha /\!\!/_{\hat{c}} \beta$ to be the array of cells

$$\alpha /\!\!/_{\hat{c}} \beta = \{(i,j) \mid (i,j) \in \alpha \text{ and } (i,j) \notin \beta\}.$$

Following the vocabulary regarding skew shapes, we refer to β as the *inner shape* and to α as the *outer shape*. The *size* of $\alpha /\!\!/_{\hat{c}} \beta$ is $|\alpha /\!\!/_{\hat{c}} \beta| = |\alpha| - |\beta|$. The skew shape $\alpha /\!\!/_{\hat{c}} \emptyset$ is the same as the Young composition diagram α. Consequently, we write $\alpha /\!\!/_{\hat{c}} \emptyset$ as α and say it is of *straight shape*.

Example 4.1.5. In this example the inner shape is denoted by cells filled with a •.

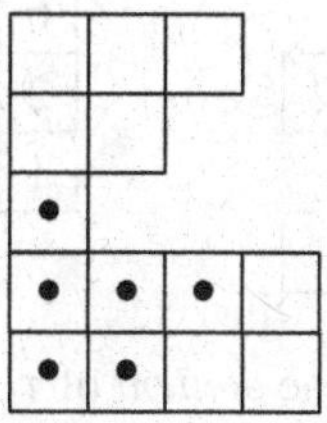

$$\alpha /\!/_{\hat{c}} \beta = (4,4,1,2,3) /\!/_{\hat{c}} (2,3,1)$$

Now we will define tableaux analogous to Young tableaux.

Definition 4.1.6. Given a skew Young composition shape $\alpha /\!/_{\hat{c}} \beta$, we define a *semistandard Young composition tableau* (abbreviated to *SSYCT*) τ of *shape* $sh(\tau) = \alpha /\!/_{\hat{c}} \beta$ to be a filling

$$\tau : \alpha /\!/_{\hat{c}} \beta \to \mathbb{Z}^+$$

of the cells of $\alpha /\!/_{\hat{c}} \beta$ such that

1. the entries in each row are weakly increasing when read from left to right
2. the entries in the first column are strictly increasing when read from the row with the smallest index to the largest index
3. set $\tau(i,j) = 0$ for all $(i,j) \in \beta$. If $i > j$ and $(j,k+1) \in \alpha /\!/_{\hat{c}} \beta$ and $\tau(i,k) \leqslant \tau(j,k+1)$, then $\tau(i,k+1) < \tau(j,k+1)$.

A *standard Young composition tableau* (abbreviated to *SYCT*) is an SSYCT in which the filling is a bijection $\tau : \alpha /\!/_{\hat{c}} \beta \to [\, |\alpha /\!/_{\hat{c}} \beta| \,]$, that is, each of the numbers $1, 2, \ldots, |\alpha /\!/_{\hat{c}} \beta|$ appears exactly once. Sometimes we will abuse notation and use SSYCTs and SYCTs to denote the set of all such tableaux.

Intuitively we can think of the third condition as saying that if $a \geqslant b$ then $a > c$ in the following subarray of cells.

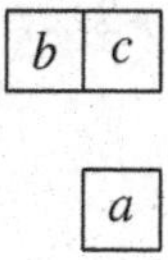

It is not hard to check that the entries within each column of either type of tableaux are distinct.

Example 4.1.7. An SSYCT and SYCT, respectively, are shown below.

7			
5	5		
•	•	2	2
•	2	3	
•	•	•	1

7	9		
6			
2	3	4	8
1	5		

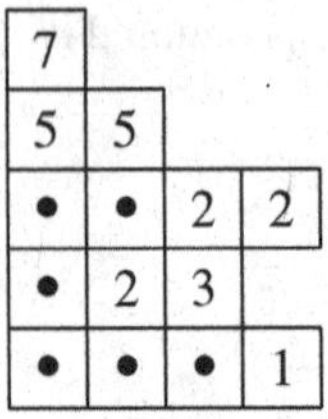

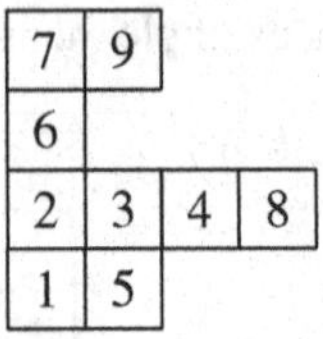

Given an SSYCT τ, we define the *content* of τ, denoted by $\mathrm{cont}(\tau)$, to be the list of nonnegative integers

$$\mathrm{cont}(\tau) = (c_1, c_2, \ldots, c_{max})$$

where c_i is the number of times i appears in τ, and *max* is the largest integer appearing in τ. Furthermore, given variables $x_1, x_2, \ldots$, we define the *monomial of τ* to be

$$x^T = x_1^{c_1} x_2^{c_2} \cdots x_{max}^{c_{max}}.$$

Given an SYCT τ, its *column reading word*, denoted by $w_{col}(\tau)$, is obtained by listing the entries from the leftmost column in *decreasing* order, followed by the entries from the second leftmost column, again in decreasing order, and so on.

The *descent set* of an SYCT τ of size n, denoted by $\mathrm{Des}(\tau)$, is the subset of $[n-1]$ consisting of all entries i of τ such that $i+1$ appears in the same column or a column to the left, that is,

$$\mathrm{Des}(\tau) = \{i \mid i+1 \text{ appears weakly left of } i\} \subseteq [n-1]$$

and the corresponding *descent composition* of τ is

$$\mathrm{comp}(\tau) = \mathrm{comp}(\mathrm{Des}(\tau)).$$

Given a composition $\alpha = (\alpha_1, \ldots, \alpha_k)$, the *canonical SYCT* U_α is the unique SYCT satisfying $sh(U_\alpha) = \alpha$ and $\mathrm{comp}(U_\alpha) = (\alpha_1, \ldots, \alpha_k)$. In U_α the first row is filled with $1, 2, \ldots, \alpha_1$ and row i for $2 \leqslant i \leqslant \ell(\alpha)$ is filled with

$$x+1, x+2, \ldots, x+\alpha_i$$

where $x = \alpha_1 + \cdots + \alpha_{i-1}$.

Example 4.1.8.

$\tau =$

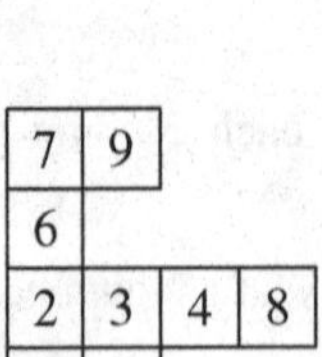

7	9		
6			
2	3	4	8
1	5		

$U_{(2,4,1,2)} =$

8	9		
7			
3	4	5	6
1	2		

$$\mathrm{Des}(\tau) = \{1, 4, 5, 6, 8\}$$
$$\mathrm{comp}(\tau) = (1, 3, 1, 1, 2, 1)$$
$$w_{col}(\tau) = 7621\ 953\ 4\ 8$$

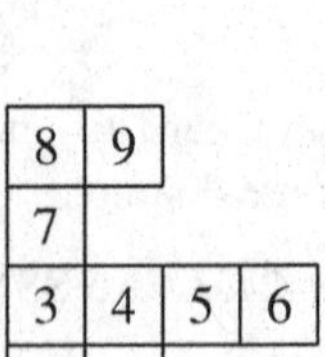

There is a bijection between SYCTs and saturated chains, whose proof is analogous to that of [15, Proposition 2.11].

Proposition 4.1.9. *A one-to-one correspondence between saturated chains in $\mathscr{L}_{\hat{c}}$ and SYCTs is given by*

$$\alpha^0 \lessdot_{\hat{c}} \alpha^1 \lessdot_{\hat{c}} \alpha^2 \lessdot_{\hat{c}} \cdots \lessdot_{\hat{c}} \alpha^n \leftrightarrow \tau$$

where τ is the SYCT of shape $\alpha^n /\!/_{\hat{c}} \alpha^0$ such that the number i appears in the cell in τ that exists in α^i but not α^{i-1}.

Example 4.1.10. In $\mathscr{L}_{\hat{c}}$ the saturated chain

$$\emptyset \lessdot_{\hat{c}} (1) \lessdot_{\hat{c}} (1,1) \lessdot_{\hat{c}} (1,2) \lessdot_{\hat{c}} (1,3) \lessdot_{\hat{c}} (2,3)$$
$$\lessdot_{\hat{c}} (2,3,1) \lessdot_{\hat{c}} (2,3,1,1) \lessdot_{\hat{c}} (2,4,1,1) \lessdot_{\hat{c}} (2,4,1,2)$$

corresponds to the following SYCT

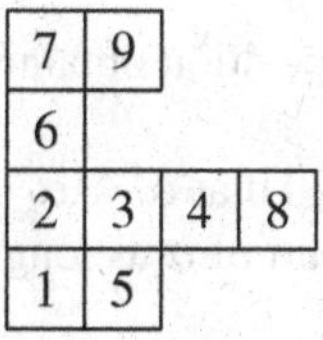

while the saturated chain in $\mathscr{L}_{\hat{c}}$

$$(1) \lessdot_{\hat{c}} (1,1) \lessdot_{\hat{c}} (1,2) \lessdot_{\hat{c}} (2,2) \lessdot_{\hat{c}} (2,3) \lessdot_{\hat{c}} (2,3,1)$$

corresponds to the following SYCT.

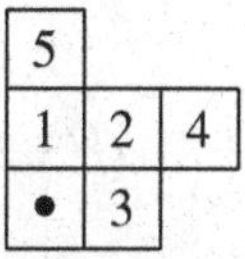

4.2 Reverse composition tableaux and the reverse composition poset

It is important to note that the combinatorial concepts used in this section are those used in [15, 40] and are related to reverse *tableaux, rather than the more common Young tableaux.*

We introduce another way to associate compositions with diagrams that differs from the association between compositions and ribbons in Section 2.3.

Definition 4.2.1. Given a composition $\alpha = (\alpha_1, \ldots, \alpha_{\ell(\alpha)}) \vDash n$, we say the *reverse composition diagram* of α, also denoted by α, is the left-justified array of n cells with α_i cells in the i-th row. We follow the Matrix or English convention, which means that we number the rows from top to bottom, and the columns from left to right. The cell in the i-th row and j-th column is denoted by the pair (i, j).

Example 4.2.2.

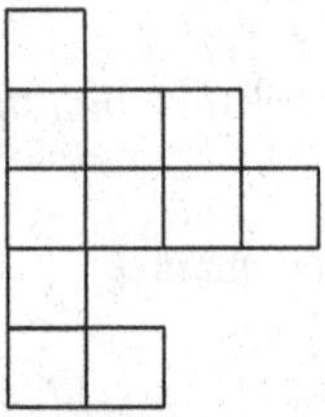

$$\alpha = (1,3,4,1,2)$$

Definition 4.2.3. The *reverse composition poset* $\mathscr{L}_{\check{c}}$ is the poset consisting of all compositions in which $\alpha = (\alpha_1, \ldots, \alpha_\ell)$ is covered by

1. $(1, \alpha_1, \ldots, \alpha_\ell)$, that is, the composition obtained by *prefixing* a part of size 1 to α.
2. $(\alpha_1, \ldots, \alpha_k + 1, \ldots, \alpha_\ell)$, provided that $\alpha_i \neq \alpha_k$ for all $i < k$, that is, the composition obtained by adding 1 to a part of α as long as that part is the *leftmost* part of that size.

As with Young's lattice, we can note the *column sequence* of a saturated chain in $\mathscr{L}_{\check{c}}$.

Example 4.2.4. A saturated chain in $\mathscr{L}_{\check{c}}$ is

$$(1) \lessdot_{\check{c}} (1,1) \lessdot_{\check{c}} (2,1) \lessdot_{\check{c}} (2,2) \lessdot_{\check{c}} (3,2) \lessdot_{\check{c}} (1,3,2)$$

and

$$\mathrm{col}(\,(1) \lessdot_{\check{c}} (1,1) \lessdot_{\check{c}} (2,1) \lessdot_{\check{c}} (2,2) \lessdot_{\check{c}} (3,2) \lessdot_{\check{c}} (1,3,2)\,) \;=\; 1,2,2,3,1.$$

Let α, β be two reverse composition diagrams such that $\beta <_{\check{c}} \alpha$. Then we define the *skew reverse composition shape* $\alpha/\!\!/_{\check{c}}\beta$ to be the array of cells

$$\alpha/\!\!/_{\check{c}}\beta = \{(i,j) \mid (i,j) \in \alpha \text{ and } (i,j) \notin \beta\}$$

where β has been drawn in the bottom left corner. As with skew shapes, we refer to β as the *inner shape* and to α as the *outer shape*. The *size* of $\alpha/\!\!/_{\check{c}}\beta$ is $|\alpha/\!\!/_{\check{c}}\beta| = |\alpha| - |\beta|$. Plus the skew shape $\alpha/\!\!/_{\check{c}}\emptyset$ is simply the reverse composition diagram α. Hence, we write α instead of $\alpha/\!\!/_{\check{c}}\emptyset$ and say it is of *straight shape*.

Example 4.2.5. In this example the inner shape is denoted by cells filled with a $\bullet$.

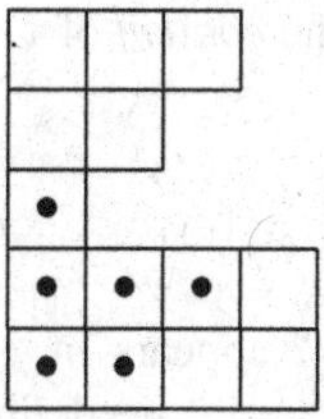

$$\alpha /\!\!/_{\check{c}} \beta = (3,2,1,4,4) /\!\!/_{\check{c}} (1,3,2)$$

Definition 4.2.6. Given a skew reverse composition shape $\alpha /\!\!/_{\check{c}} \beta$, we define a *semistandard reverse composition tableau* (abbreviated to *SSRCT*) $\check{\tau}$ of *shape* $sh(\check{\tau}) = \alpha /\!\!/_{\check{c}} \beta$ to be a filling

$$\check{\tau} : \alpha /\!\!/_{\check{c}} \beta \to \mathbb{Z}^{+}$$

of the cells of $\alpha /\!\!/_{\check{c}} \beta$ such that

1. the entries in each row are weakly decreasing when read from left to right
2. the entries in the first column are strictly increasing when read from the row with the smallest index to the largest index
3. set $\check{\tau}(i,j) = \infty$ for all $(i,j) \in \beta$. If $i < j$ and $(j,k+1) \in \alpha /\!\!/_{\check{c}} \beta$ and $\check{\tau}(i,k) \geqslant \check{\tau}(j,k+1)$, then $\check{\tau}(i,k+1) > \check{\tau}(j,k+1)$.

A *standard reverse composition tableau* (abbreviated to *SRCT*) is an SSRCT in which the filling is a bijection $\check{\tau} : \alpha /\!\!/_{\check{c}} \beta \to [\,|\alpha /\!\!/_{\check{c}} \beta|\,]$, that is, each of the numbers $1, 2, \ldots, |\alpha /\!\!/_{\check{c}} \beta|$ appears exactly once. Sometimes we will abuse notation and use SSRCTs and SRCTs to denote the set of all such tableaux.

Intuitively we can think of the third condition as saying that if $a \leqslant b$ then $a < c$ in the following subarray of cells.

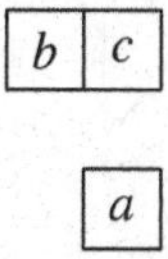

It is not hard to check that the entries within each column of either type of tableaux are distinct.

Example 4.2.7. An SSRCT and SRCT, respectively, are shown below.

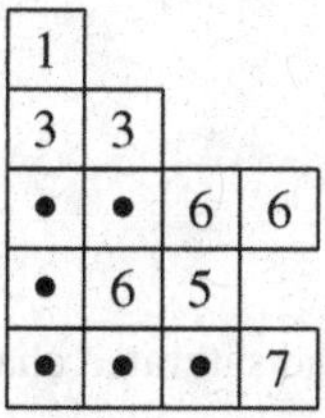

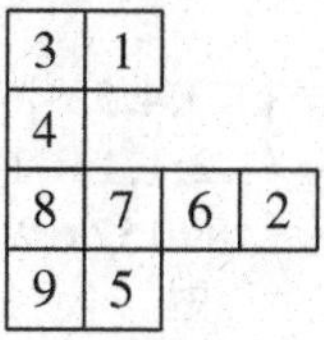

Given an SSRCT $\check{\tau}$, we define the *content* of $\check{\tau}$, denoted by $\mathrm{cont}(\check{\tau})$, to be the list of nonnegative integers

$$\mathrm{cont}(\check{\tau}) = (c_1, c_2, \ldots, c_{max})$$

where c_i is the number of times i appears in $\check{\tau}$, and max is the largest integer appearing in $\check{\tau}$. Given variables $x_1, x_2, \ldots$, we define the *monomial of* $\check{\tau}$ to be

$$x^{\check{\tau}} = x_1^{c_1} x_2^{c_2} \cdots x_{max}^{c_{max}}.$$

Given an SRCT $\check{\tau}$, its *column reading word*, denoted by $w_{col}(\check{\tau})$, is obtained by listing the entries from the leftmost column in *increasing* order, followed by the entries from the second leftmost column, again in increasing order, and so on.

The *descent set* of an SRCT $\check{\tau}$ of size n, denoted by $\mathrm{Des}(\check{\tau})$, is the subset of $[n-1]$ consisting of all entries i of $\check{\tau}$ such that $i+1$ appears in the same column or a column to the right, that is,

$$\mathrm{Des}(\check{\tau}) = \{i \mid i+1 \text{ appears weakly right of } i\} \subseteq [n-1]$$

and the corresponding *descent composition* of $\check{\tau}$ is

$$\mathrm{comp}(\check{\tau}) = \mathrm{comp}(\mathrm{Des}(\check{\tau})).$$

Given a composition $\alpha = (\alpha_1, \ldots, \alpha_k)$, the *canonical* SRCT $\check{U}_\alpha$ is the unique SRCT satisfying $sh(\check{U}_\alpha) = \alpha$ and $\mathrm{comp}(\check{U}_\alpha) = (\alpha_1, \ldots, \alpha_k)$. In $\check{U}_\alpha$ the first row is filled with $1, 2, \ldots, \alpha_1$ and row i for $2 \leqslant i \leqslant \ell(\alpha)$ is filled with

$$x+1, x+2, \ldots, x+\alpha_i$$

where $x = \alpha_1 + \cdots + \alpha_{i-1}$.

Example 4.2.8.

$\check{\tau} =$

3	1		
4			
8	7	6	2
9	5		

$\check{U}_{(2,1,4,2)} =$

2	1		
3			
7	6	5	4
9	8		

$$\mathrm{Des}(\check{\tau}) = \{1,3,4,5,8\}$$
$$\mathrm{comp}(\check{\tau}) = (1,2,1,1,3,1)$$
$$w_{col}(\check{\tau}) = 3489\ 157\ 6\ 2$$

Again there is a bijection between SRCTs and saturated chains in $\mathscr{L}_{\check{\mathcal{C}}}$.

Proposition 4.2.9. *[15, Proposition 2.11] A one-to-one correspondence between saturated chains in $\mathcal{L}_{\check{c}}$ and SRCTs is given by*

$$\alpha^0 \lessdot_{\check{c}} \alpha^1 \lessdot_{\check{c}} \alpha^2 \lessdot_{\check{c}} \cdots \lessdot_{\check{c}} \alpha^n \leftrightarrow \check{\tau}$$

where $\check{\tau}$ is the SRCT of shape $\alpha^n /\!/_{\check{c}} \alpha^0$ such that the number $n-i+1$ appears in the cell in $\check{\tau}$ that exists in α^i but not α^{i-1}.

Example 4.2.10. The saturated chain in $\mathcal{L}_{\check{c}}$

$$\emptyset \lessdot_{\check{c}} (1) \lessdot_{\check{c}} (1,1) \lessdot_{\check{c}} (2,1) \lessdot_{\check{c}} (3,1) \lessdot_{\check{c}} (3,2)$$
$$\lessdot_{\check{c}} (1,3,2) \lessdot_{\check{c}} (1,1,3,2) \lessdot_{\check{c}} (1,1,4,2) \lessdot_{\check{c}} (2,1,4,2)$$

corresponds to the following SRCT.

3	1		
4			
8	7	6	2
9	5		

Meanwhile the saturated chain in $\mathcal{L}_{\check{c}}$

$$(1) \lessdot_{\check{c}} (1,1) \lessdot_{\check{c}} (2,1) \lessdot_{\check{c}} (2,2) \lessdot_{\check{c}} (3,2) \lessdot_{\check{c}} (1,3,2)$$

corresponds to the following SRCT.

1		
5	4	2
•	3	

4.3 Bijections between composition tableaux and other tableaux

Note that the map $\Gamma : \mathcal{L}_{\hat{c}} \to \mathcal{L}_{\check{c}}$ defined by

$$\Gamma(\alpha) = \alpha^r$$

is an isomorphism of graded posets. Let $\mathscr{C}(P)$ denote the set of all saturated chains of finite length in the poset P. If P and Q are graded posets, then we say that a map $\varphi : \mathscr{C}(P) \to \mathscr{C}(Q)$ is *rank-preserving* if given a chain $C \in \mathscr{C}(P)$ of length k, $\varphi(C)$

also has length k and the corresponding elements of C and $\varphi(C)$ have the same rank. Then Γ induces a rank-preserving bijection between $\mathscr{C}(\mathscr{L}_{\hat{c}})$ and $\mathscr{C}(\mathscr{L}_{\check{c}})$, namely,

$$(\alpha^1 \lessdot_{\hat{c}} \cdots \lessdot_{\hat{c}} \alpha^k) \overset{\Gamma}{\longmapsto} (\Gamma(\alpha^1) \lessdot_{\check{c}} \cdots \lessdot_{\check{c}} \Gamma(\alpha^k)). \tag{4.1}$$

Note that by Propositions 4.1.9 and 4.2.9, Γ induces a shape-reversing bijection between SYCTs and SRCTs that, for an SYCT with n cells, replaces each entry i by the entry $n-i+1$.

There also exists a bijection between SSRTs and SSRCTs, which was introduced in [64] along with certain properties, such as commuting with RSK [64, Proposition 3.3], and generalized in [15, Proposition 2.17]. It is the generalization that we now recall, denoting it by $\check{\rho}$.

Let $SSRCT(-/\!\!/_{\check{c}}\alpha)$ denote the set of all SSRCTs with inner shape α, and let $SSRT(-/\widetilde{\alpha})$ denote the set of all SSRTs with inner shape $\widetilde{\alpha}$. Then

$$\check{\rho}_\alpha : SSRCT(-/\!\!/_{\check{c}}\alpha) \to SSRT(-/\widetilde{\alpha}) \tag{4.2}$$

is defined as follows. Given an SSRCT $\check{\tau}$, obtain $\check{\rho}_\alpha(\check{\tau})$ by writing the entries in each column in decreasing order and bottom justifying these new columns on the inner shape $\widetilde{\alpha}$ if it exists. Note that by definition, if given an SRCT $\check{\tau}$ with $sh(\check{\tau}) = \alpha/\!\!/_{\check{c}}\beta$, then $\mathrm{Des}(\check{\tau}) = \mathrm{Des}(\check{\rho}_\beta(\check{\tau}))$.

Example 4.3.1.

2	2	2	2	
4	3			
7	6	5	4	3
8	5	4		

$\xrightarrow{\check{\rho}_\emptyset}$

2	2			
4	3	2		
7	5	4	2	
8	6	5	4	3

The inverse map

$$\check{\rho}_\alpha^{-1} : SSRT(-/\widetilde{\alpha}) \to SSRCT(-/\!\!/_{\check{c}}\alpha) \tag{4.3}$$

is also straightforward to define.

Given an SSRT $\check{T}$:

1. Take the set of i entries in the first column of $\check{T}$ and write them in increasing order in rows $1,2,\ldots,i$ above the inner shape that has a cell in position $(i+1,1)$ but not $(i,1)$ to form the first column of $\check{\tau}$.
2. Take the set of entries in column 2 in decreasing order and place them in the row with the smallest index so that either
   - the cell to the immediate left of the number being placed is filled and the row entries weakly decrease when read from left to right
   - the cell to the immediate left of the number being placed belongs to the inner shape.
3. Repeat the previous step with the set of entries in column k for $k = 3,\ldots,\widetilde{\alpha}_1$.

Example 4.3.2.

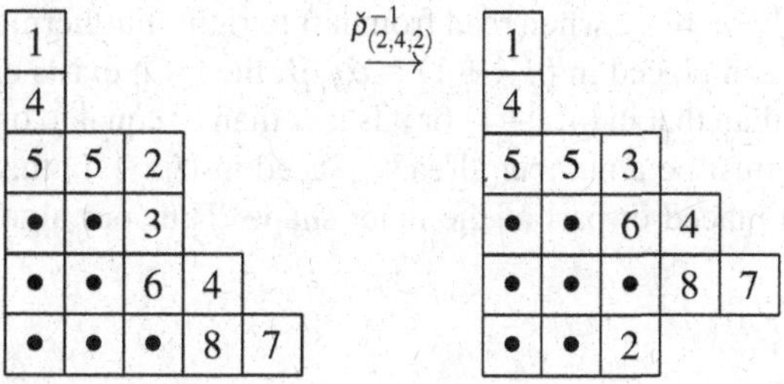

Analogously, there exists a bijection between SSYTs and SSYCTs that is new to the literature. Now let $SSYCT(-/\!\!/_{\hat{c}}\alpha)$ denote the set of all SSYCTs with inner shape α, and $SSYT(-/\widetilde{\alpha})$ denote the set of all SSYTs with inner shape $\widetilde{\alpha}$. Then

$$\hat{\rho}_\alpha : SSYCT(-/\!\!/_{\hat{c}}\alpha) \to SSYT(-/\widetilde{\alpha}) \qquad (4.4)$$

is defined on an SSYCT τ to be the SSYT $\hat{\rho}(\tau)$ obtained by writing the entries in each column of τ in increasing order and bottom justifying these new columns on the inner shape $\widetilde{\alpha}$ if it exists. Note that by definition, if given an SYCT τ with $sh(\tau) = \alpha/\!\!/_{\hat{c}}\beta$, then $\mathrm{Des}(\tau) = \mathrm{Des}(\hat{\rho}_\beta(\tau))$.

Example 4.3.3.

8		
5	7	10
2	3	6	9
1	4	11

$\xrightarrow{\hat{\rho}_0}$

8		
5	7	11
2	4	10
1	3	6	9

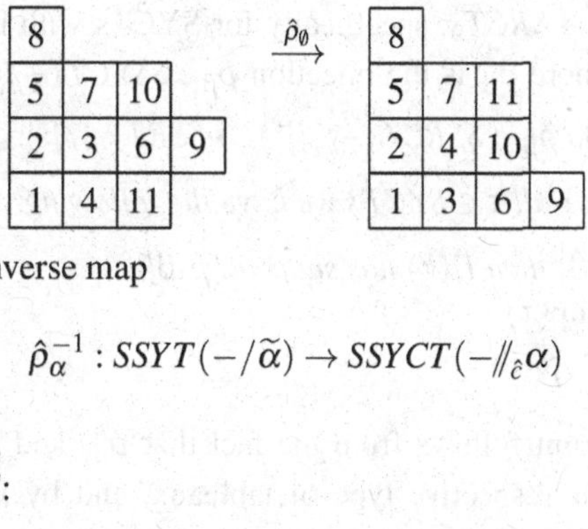

Meanwhile, the inverse map

$$\hat{\rho}_\alpha^{-1} : SSYT(-/\widetilde{\alpha}) \to SSYCT(-/\!\!/_{\hat{c}}\alpha) \qquad (4.5)$$

is defined as follows.

Given an SSYT T:

1. If the first column of the inner shape has i cells, then take the set of entries in the first column of T and write them in increasing order in rows $i+1, i+2, \ldots$ to form the first column of τ.
2. Take the set of entries in column 2 in increasing order and place them in the row with the largest index so that either

   - the cell to the immediate left of the number being placed is filled and the row entries weakly increase when read from left to right
   - the cell to the immediate left of the number being placed belongs to the inner shape.

3. Repeat the previous step with the set of entries in column k for $k = 3, \ldots, \widetilde{\alpha}_1$.

Note that by construction the entries in the first column are strictly increasing when read from the row with the smallest index to the largest index, and the entries in each row weakly increase when read from left to right. Furthermore, if $i > j$ and a number has just been placed in $(j, k+1) \in \alpha /\!/_{\hat{c}} \beta$, then if it exists either the number in (i,k) is greater than that in $(j,k+1)$ or it is less than or equal to or part of the inner shape, and there must be a number already placed in $(i,k+1)$ that is less than the number just been placed or part of the inner shape. Thus our algorithm constructs an SSYCT.

Example 4.3.4.

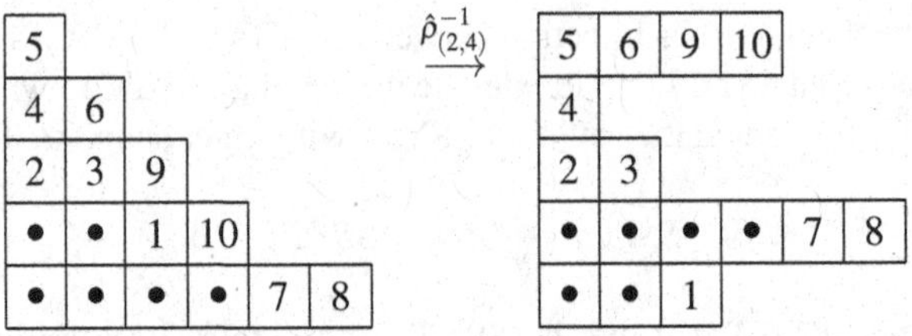

Let $\hat{\Upsilon}$ denote the bijection in Proposition 4.1.9 from saturated chains in $\mathscr{L}_{\hat{c}}$ to the set of all SYCTs, and $\check{\Upsilon}$ denote the bijection in Proposition 4.2.9 from saturated chains in $\mathscr{L}_{\check{c}}$ to the set of all SRCTs. The bijection $\check{\Gamma} : SYTs \to SRTs$, which replaces each entry i in a tableau with n cells with $n-i+1$, induces an inner shape reversing bijection $\hat{\Gamma} : SYCTs \to SRCTs$, specifically for SYCTs with inner shape β define $\hat{\Gamma} = \check{\rho}_{\beta^r}^{-1} \circ \check{\Gamma} \circ \hat{\rho}_\beta$, where $\hat{\rho}_\beta$ is the bijection $\hat{\rho}_\beta : SSYCT(-/\!/_{\hat{c}}\beta) \to SSYT(-/\widetilde{\beta})$, and $\check{\rho}_{\beta^r}$ is the bijection $\check{\rho}_{\beta^r} : SSRCT(-/\!/_{\check{c}}\beta^r) \to SSRT(-/\widetilde{\beta})$.

Proposition 4.3.5. *For all $\tau \in SYCTs$ we have the following.*

1. *If τ has shape $\alpha /\!/_{\hat{c}} \beta$, then $\hat{\Gamma}(\tau)$ has shape $\alpha^r /\!/_{\check{c}} \beta^r$.*
2. $\mathrm{comp}(\hat{\Gamma}(\tau)) = \mathrm{comp}(\tau)^r$.

Moreover, $\hat{\Gamma} = \check{\Upsilon} \circ \Gamma \circ \hat{\Upsilon}^{-1}$.

Proof. The second point follows from the fact that $\check{\rho}_{\beta^r}$ and $\hat{\rho}_\beta$ preserve descent compositions for their respective type of tableaux, and by Proposition 2.5.6, $\check{\Gamma}$ reverses descent compositions.

The claim that $\hat{\Gamma} = \check{\Upsilon} \circ \Gamma \circ \hat{\Upsilon}^{-1}$ follows from several facts. First, Γ preserves the column sequence of each chain, giving a correspondence between cells. The fact that $\Gamma(\alpha) = \alpha^r$ reflects the indexing convention used for the composition shapes of SYCT versus SRCT. Lastly, $\hat{\Upsilon}$ assigns the entry i to the cell introduced in the i-th cover relation, while $\check{\Upsilon}$ assigns the entry $n-i+1$ to the corresponding cell. Thus the set of entries in the corresponding columns of $\check{\Upsilon}(\Gamma(\hat{\Upsilon}^{-1}(\tau)))$ are precisely those of $\hat{\Gamma}(\tau)$. But the set of entries in each column completely characterizes each tableau. Thus $\check{\Upsilon}(\Gamma(\hat{\Upsilon}^{-1}(\tau))) = \hat{\Gamma}(\tau)$.

Finally, the first point follows from the fact that Γ is the reversal of compositions, that is, that $\check{\Upsilon} \circ \Gamma \circ \hat{\Upsilon}^{-1}$ maps an SYCT of shape $\alpha /\!/_{\hat{c}} \beta$ to an SRCT of shape $\alpha^r /\!/_{\check{c}} \beta^r$. □

Chapter 5
Quasisymmetric Schur functions

Abstract In this final chapter we introduce two additional bases for the Hopf algebra of quasisymmetric functions. The first is the basis of quasisymmetric Schur functions already in the literature, whose combinatorics is connected to reverse composition tableaux. The second is the new basis of Young quasisymmetric Schur functions whose combinatorics is connected to Young composition tableaux. For each of these bases we determine their expansion in terms of fundamental quasisymmetric functions, monomial quasisymmetric functions and monomials, and see how they refine Schur functions in a natural way. We then, for each basis, describe Pieri rules and define skew analogues, consequently developing a Littlewood-Richardson rule for these skew analogues and the coproduct. Finally via duality, we introduce two new bases for the Hopf algebra of noncommutative symmetric functions, each of which projects onto the basis of Schur functions under the forgetful map. Each of these new bases exhibit Pieri and Littlewood-Richardson rules, which we describe. As with their quasisymmetric counterparts, one basis involves reverse composition tableaux, while the other involves Young composition tableaux.

5.1 Original quasisymmetric Schur functions

We now arrive at quasisymmetric Schur functions, which we will initially choose to define in terms of fundamental quasisymmetric functions in analogy with Equation (3.18).

Definition 5.1.1. [40, Theorem 6.2] Let α be a composition. Then the *quasisymmetric Schur function* $\check{\mathscr{S}}_\alpha$ is defined by $\check{\mathscr{S}}_\emptyset = 1$ and

$$\check{\mathscr{S}}_\alpha = \sum_\beta \check{d}_{\alpha\beta} F_\beta$$

K. Luoto et al., *An Introduction to Quasisymmetric Schur Functions*,
SpringerBriefs in Mathematics, DOI 10.1007/978-1-4614-7300-8_5,
© Kurt Luoto, Stefan Mykytiuk, Stephanie van Willigenburg 2013

where the sum is over all compositions $\beta \vDash |\alpha|$ and $\check{d}_{\alpha\beta}$ = the number of *SRCTs* $\check{\tau}$ of shape α such that $\mathrm{Des}(\check{\tau}) = \mathrm{set}(\beta)$.

Example 5.1.2. We have $\check{\mathcal{S}}_{(3,2)} = F_{(3,2)} + F_{(2,2,1)} + F_{(1,3,1)}$ from the SRCTs

$$\begin{array}{|c|c|c|}\hline 3 & 2 & 1 \\ \hline 5 & 4 & \multicolumn{1}{c}{} \\ \cline{1-2}\end{array} \qquad \begin{array}{|c|c|c|}\hline 4 & 3 & 1 \\ \hline 5 & 2 & \multicolumn{1}{c}{} \\ \cline{1-2}\end{array} \qquad \begin{array}{|c|c|c|}\hline 4 & 3 & 2 \\ \hline 5 & 1 & \multicolumn{1}{c}{} \\ \cline{1-2}\end{array}$$

and $\check{\mathcal{S}}_{(2,3)} = F_{(2,3)} + F_{(1,2,2)}$ from the following SRCTs.

$$\begin{array}{|c|c|c|}\cline{1-2} 2 & 1 & \multicolumn{1}{c}{} \\ \hline 5 & 4 & 3 \\ \hline\end{array} \qquad \begin{array}{|c|c|c|}\cline{1-2} 3 & 1 & \multicolumn{1}{c}{} \\ \hline 5 & 4 & 2 \\ \hline\end{array}$$

Quasisymmetric Schur functions also have an expansion in the basis of monomial quasisymmetric functions that is analogous to the expansion of Schur functions in the basis of monomial symmetric functions discussed in Proposition 3.2.10.

Proposition 5.1.3. *[40, Theorem 6.1] Let $\alpha \vDash n$. Then*

$$\check{\mathcal{S}}_\alpha = \sum_{\beta \vDash n} \check{K}_{\alpha\beta} M_\beta$$

where $\check{\mathcal{S}}_\emptyset = 1$ and $\check{K}_{\alpha\beta}$ is the number of SSRCTs $\check{\tau}$ satisfying $sh(\check{\tau}) = \alpha$ and $\mathrm{cont}(\check{\tau}) = \beta$.

Example 5.1.4. We have $\check{\mathcal{S}}_{(1,2)} = M_{(1,2)} + M_{(1,1,1)}$ from the following SSRCTs.

$$\begin{array}{|c|c|}\cline{1-1} 1 & \multicolumn{1}{c}{} \\ \hline 2 & 2 \\ \hline\end{array} \qquad \begin{array}{|c|c|}\cline{1-1} 1 & \multicolumn{1}{c}{} \\ \hline 3 & 2 \\ \hline\end{array}$$

As with Schur functions, quasisymmetric Schur functions can also be described as a sum of monomials arising from tableaux, analogous to Definition 3.2.8.

Proposition 5.1.5. *[40, Definition 5.1] Let α be a composition. Then $\check{\mathcal{S}}_\emptyset = 1$ and*

$$\check{\mathcal{S}}_\alpha = \sum_{\check{\tau}} x^{\check{\tau}}$$

where the sum is over all SSRCTs $\check{\tau}$ of shape α.

In [40, Proposition 5.5] it was shown that the set of all quasisymmetric Schur functions forms a $\mathbb{Z}$-basis for QSym and

$$\mathrm{QSym}^n = \mathrm{span}\{\check{\mathcal{S}}_\alpha \mid \alpha \vDash n\}.$$

Furthermore, it was shown

$$s_\lambda = \sum_{\tilde{\alpha}=\lambda} \check{\mathscr{S}}_\alpha, \tag{5.1}$$

which is identical to the relationship between the monomial symmetric functions and monomial quasisymmetric functions given in Equation (3.17):

$$m_\lambda = \sum_{\tilde{\alpha}=\lambda} M_\alpha$$

leading to the natural naming of quasisymmetric Schur functions.

5.2 Young quasisymmetric Schur functions

The basis of Schur functions in Sym is preserved under the involutions ψ, ω, ρ in Section 3.6. However, the basis of quasisymmetric Schur functions in QSym is not. The image of this basis under ω is the basis of row-strict quasisymmetric Schur functions investigated in [29, 65]. As we will see, applying the automorphism ρ : QSym $\to$ QSym, given by

$$\rho(F_\alpha) = F_{\alpha^r},$$

to the basis of quasisymmetric Schur functions will yield a new basis for QSym whose combinatorics is based on Young composition tableaux, and will give rise to theorems analogous to those theorems in symmetric function theory involving Young tableaux. Properties of quasisymmetric Schur functions under $\psi = \rho \circ \omega = \omega \circ \rho$ would involve row-strict Young composition tableaux.

Definition 5.2.1. Let α be a composition. Then the *Young quasisymmetric Schur function* $\hat{\mathscr{S}}_\alpha$ is defined by

$$\hat{\mathscr{S}}_\alpha = \rho(\check{\mathscr{S}}_{\alpha^r}).$$

Expanding these new functions in terms of fundamental quasisymmetric functions gives us another expression analogous to Equation (3.18).

Proposition 5.2.2. *Let α be a composition. Then $\hat{\mathscr{S}}_\emptyset = 1$ and*

$$\hat{\mathscr{S}}_\alpha = \sum_\beta \hat{d}_{\alpha\beta} F_\beta$$

where the sum is over all compositions $\beta \models |\alpha|$ and $\hat{d}_{\alpha\beta} =$ the number of SYCTs τ of shape α such that $\mathrm{Des}(\tau) = \mathrm{set}(\beta)$.

Proof. Note from Definition 5.1.1 that

$$\check{\mathscr{S}}_{\alpha^r} = \sum_\beta \check{d}_{\alpha^r\beta^r} F_{\beta^r}$$

where

$$\check{d}_{\alpha^r \beta^r} = \#\{\check{\tau} \in SRCTs \mid sh(\check{\tau}) = \alpha^r, \mathrm{Des}(\check{\tau}) = \mathrm{set}(\beta^r)\}.$$

Thus,

$$\hat{\mathscr{S}}_\alpha = \rho(\check{\mathscr{S}}_{\alpha^r}) = \sum_\beta \check{d}_{\alpha^r \beta^r} F_\beta.$$

However,

$$\check{d}_{\alpha^r \beta^r} = \#\{\tau \in SYCTs \mid sh(\tau) = \alpha, \mathrm{Des}(\tau) = \mathrm{set}(\beta)\} = \hat{d}_{\alpha\beta}$$

since by Proposition 4.3.5 the bijection $\hat{\Gamma} : SYCTs \to SRCTs$ reverses both indexing shape and descent set. □

Example 5.2.3. We have $\hat{\mathscr{S}}_{(3,2)} = F_{(3,2)} + F_{(2,2,1)}$ from the SYCTs

4	5	
1	2	3

3	5	
1	2	4

and $\hat{\mathscr{S}}_{(2,3)} = F_{(2,3)} + F_{(1,2,2)} + F_{(1,3,1)}$ from the following SYCTs.

3	4	5
1	2	

2	3	5
1	4	

2	3	4
1	5	

An analogue of Proposition 3.2.10 also exists.

Proposition 5.2.4. *Let $\alpha \vDash n$. Then*

$$\hat{\mathscr{S}}_\alpha = \sum_{\beta \vDash n} \hat{K}_{\alpha\beta} M_\beta$$

where $\hat{\mathscr{S}}_\emptyset = 1$ and $\hat{K}_{\alpha\beta}$ is the number of SSYCTs τ satisfying $sh(\tau) = \alpha$ and $\mathrm{cont}(\tau) = \beta$.

Proof. Note that Proposition 5.2.2 can equivalently be written

$$\hat{\mathscr{S}}_\alpha = \sum_\tau F_{\mathrm{comp}(\tau)}$$

where the sum is over all SYCTs τ of shape α. Since $F_\beta = \sum_{\beta \succcurlyeq \gamma} M_\gamma$, if $\beta = (\beta_1, \ldots, \beta_k)$ and $\gamma = (\gamma_1, \ldots, \gamma_{\ell_k})$, then we know any leading term of M_γ

$$x_1^{\gamma_1} x_2^{\gamma_2} \cdots x_{\ell_k}^{\gamma_{\ell_k}}$$

appearing in the sum has exponents satisfying the following.

$$\gamma_1+\gamma_2+\cdots+\gamma_{\ell_1}=\beta_1$$
$$\gamma_{\ell_1+1}+\gamma_{\ell_1+2}+\cdots+\gamma_{\ell_2}=\beta_2$$
$$\vdots$$
$$\gamma_{\ell_{k-1}+1}+\gamma_{\ell_{k-1}+2}+\cdots+\gamma_{\ell_k}=\beta_k$$

Given an SYCT τ with $sh(\tau)=\alpha$ and $\mathrm{comp}(\tau)=\beta$, form τ' by applying the map

$$1,\ldots,\gamma_1 \mapsto 1$$
$$\left(\sum_{r=1}^{m-1}\gamma_r\right)+1,\ldots,\sum_{r=1}^{m}\gamma_r \mapsto m \text{ for } m=2,\ldots,\ell_k.$$

Then τ' is an SSYCT since τ is an SYCT, so under the map the entries in the rows of τ' weakly increase when read from left to right, and increase along the first column. Plus if there exists $i>j$ so $\tau'(i,k)\leqslant\tau'(j,k+1)$ but $\tau'(i,k+1)\geqslant\tau'(j,k+1)$ or $(i,k+1)$ is not a cell in τ', then in τ we must have had $\tau(i,k)<\tau(j,k+1)$ and $\tau(i,k+1)>\tau(j,k+1)$ or $(i,k+1)$ is not a cell in τ, which is a contradiction. Note that this process is reversible. Thus the coefficient of M_γ in the expansion of $\hat{\mathscr{S}}_\alpha$ is the number of SSYCTs τ satisfying $sh(\tau)=\alpha$ and $\mathrm{cont}(\tau)=\gamma$. □

Example 5.2.5. We have $\hat{\mathscr{S}}_{(1,2)}=M_{(1,2)}+M_{(1,1,1)}$ from the following SSYCTs.

2	2
1	

2	3
1	

In analogy with Definition 3.2.8 we also have a decomposition in terms of monomials, arising from Young composition tableaux.

Proposition 5.2.6. *Let α be a composition. Then $\hat{\mathscr{S}}_\emptyset=1$ and*

$$\hat{\mathscr{S}}_\alpha=\sum_\tau x^\tau$$

where the sum is over all SSYCTs τ of shape α.

Proof. If $\beta=(\beta_1,\ldots,\beta_k)$ is a composition, then

$$M_\beta=\sum_{i_1<\cdots<i_k} x_{i_1}^{\beta_1}\cdots x_{i_k}^{\beta_k}$$

so it is sufficient to show that

1. given an SSYCT τ with $sh(\tau)=\alpha$ and $\mathrm{cont}(\tau)=(\beta_1,\ldots,\beta_k)=\beta$, that is, with β_1 ones, β_2 twos, ... , β_k ks, we can create an SSYCT τ' with $sh(\tau')=\alpha$ and β_1 i_1s, β_2 i_2s, ... , β_k i_ks, where $i_1<\cdots<i_k$; and
2. given an SSYCT τ' with $sh(\tau')=\alpha$ and β_1 i_1s, β_2 i_2s, ..., β_k i_ks, where $i_1<\cdots<i_k$, we can create an SSYCT τ with $sh(\tau)=\alpha$ and $\mathrm{cont}(\tau)=(\beta_1,\ldots,\beta_k)=\beta$, that is, with β_1 ones, β_2 twos, ... , β_k ks.

The result will then follow by Proposition 5.2.4. The first part follows by applying the map $j\mapsto i_j$ for $j=1,\ldots,k$ to every entry of τ, and the second part follows by applying the map $i_j\mapsto j$ for $j=1,\ldots,k$ to every entry of τ'. Since both maps maintain the relative order of the entries, all conditions for an SSYCT are still satisfied after each map is applied. □

Since ρ is an automorphism on QSym we have, as with quasisymmetric Schur functions, that the set of all Young quasisymmetric Schur functions forms a $\mathbb{Z}$-basis for QSym and

$$\mathrm{QSym}^n=\mathrm{span}\{\hat{\mathscr{S}}_\alpha \mid \alpha\vDash n\}.$$

Plus, since ρ restricts to the identity on Sym we have

$$s_\lambda=\sum_{\tilde{\alpha}=\lambda}\hat{\mathscr{S}}_\alpha. \tag{5.2}$$

5.3 Pieri and Littlewood-Richardson rules in QSym using reverse composition tableaux

Note that from Theorem 3.2.17, the Pieri rules for symmetric functions can be stated as follows.

Proposition 5.3.1 (Pieri rules for Schur functions). *Let λ be a partition. Then*

$$s_{(n)}s_\lambda=\sum_\mu s_\mu$$

where the sum is taken over all partitions μ such that

1. *$\delta=\mu/\lambda$ is a horizontal strip,*
2. *$|\delta|=n$.*

Also,

$$s_{(1^n)}s_\lambda=\sum_\mu s_\mu$$

where the sum is taken over all partitions μ such that

1. *$\varepsilon=\mu/\lambda$ is a vertical strip,*
2. *$|\varepsilon|=n$.*

In order for us to state analogous Pieri rules for quasisymmetric Schur functions, we need three new operators: $\check{\mathfrak{d}}, \check{\mathfrak{h}}, \check{\mathfrak{v}}$. In practice the decrementing $\check{\mathfrak{d}}_s$ operator subtracts 1 from the rightmost part of size s in a composition, or returns the empty composition. Meanwhile the $\check{\mathfrak{h}}_{\{s_1<\cdots<s_j\}}$ operator subtracts 1 from the rightmost part of size $s_j, s_{j-1}, \ldots$ recursively. Similarly, the $\check{\mathfrak{v}}_{\{m_1\leqslant\cdots\leqslant m_j\}}$ operator subtracts 1 from the rightmost part of size $m_1, m_2, \ldots$ recursively.

Example 5.3.2. If $\alpha = (1,2,3)$, then

$$\check{\mathfrak{h}}_{\{2,3\}}(\alpha) = \check{\mathfrak{d}}_2(\check{\mathfrak{d}}_3((1,2,3))) = \check{\mathfrak{d}}_2((1,2,2)) = (1,2,1)$$

and

$$\check{\mathfrak{v}}_{\{2,3\}}(\alpha) = \check{\mathfrak{d}}_3(\check{\mathfrak{d}}_2((1,2,3))) = \check{\mathfrak{d}}_3((1,1,3)) = (1,1,2).$$

For completeness we define these three operators formally. Let $\alpha = (\alpha_1, \ldots, \alpha_k)$ be a composition whose largest part is m, and let $s \in [m]$. If there exists $1 \leqslant i \leqslant k$ such that $s = \alpha_i$ and $s \neq \alpha_j$ for all $j > i$, then define

$$\check{\mathfrak{d}}_s(\alpha) = (\alpha_1, \ldots, \alpha_{i-1}, (s-1), \alpha_{i+1}, \ldots, \alpha_k),$$

otherwise define $\check{\mathfrak{d}}_s(\alpha)$ to be the empty composition. Let $S = \{s_1 < \cdots < s_j\}$. Then define

$$\check{\mathfrak{h}}_S(\alpha) = \check{\mathfrak{d}}_{s_1}(\ldots(\check{\mathfrak{d}}_{s_{j-1}}(\check{\mathfrak{d}}_{s_j}(\alpha)))\ldots).$$

Similarly let $M = \{m_1 \leqslant \cdots \leqslant m_j\}$. Then define

$$\check{\mathfrak{v}}_M(\alpha) = \check{\mathfrak{d}}_{m_j}(\ldots(\check{\mathfrak{d}}_{m_2}(\check{\mathfrak{d}}_{m_1}(\alpha)))\ldots).$$

We remove any zeros from $\check{\mathfrak{h}}_S(\alpha)$ or $\check{\mathfrak{v}}_M(\alpha)$ to obtain a composition if needs be.

For any horizontal strip δ we denote by $S(\delta)$ the set of columns its skew diagram occupies, and for any vertical strip ε we denote by $M(\varepsilon)$ the multiset of columns its skew diagram occupies, where multiplicities for a column are given by the number of cells in that column, and column indices are listed in weakly increasing order. We are now ready to state our analogous Pieri rule.

Theorem 5.3.3 (Pieri rules for quasisymmetric Schur functions). *[40, Theorem 6.3]Let α be a composition. Then*

$$\check{\mathscr{S}}_{(n)}\check{\mathscr{S}}_\alpha = \sum_\beta \check{\mathscr{S}}_\beta$$

where the sum is taken over all compositions β such that

1. $\delta = \widetilde{\beta}/\widetilde{\alpha}$ *is a horizontal strip,*
2. $|\delta| = n$,
3. $\check{\mathfrak{h}}_{S(\delta)}(\beta) = \alpha$.

Also,

$$\check{\mathscr{S}}_{(1^n)}\check{\mathscr{S}}_{\alpha}=\sum_{\beta}\check{\mathscr{S}}_{\beta}$$

where the sum is taken over all compositions β such that

1. $\varepsilon=\widetilde{\beta}/\widetilde{\alpha}$ *is a vertical strip,*
2. $|\varepsilon|=n$,
3. $\check{\mathfrak{v}}_{M(\varepsilon)}(\beta)=\alpha$.

For a more visual interpretation of Theorem 5.3.3 we use reverse composition diagrams in place of compositions in the next example. Then $\check{\mathfrak{d}}_s$ is the operation that removes the rightmost cell from the lowest row of length s.

Example 5.3.4. If we place • in the cell to be removed, then

$$\check{\mathfrak{d}}_1((1,1,3))=$$
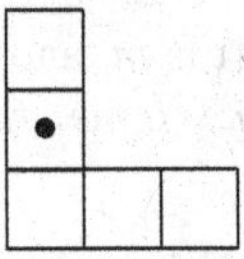
$$=(1,3).$$

If we wish to compute $\check{\mathscr{S}}_{(1)}\check{\mathscr{S}}_{(1,3)}$, then we consider the four skew diagrams

$$(4,1)/(3,1),\ (3,2)/(3,1),\ (3,1,1)/(3,1),\ (3,1,1)/(3,1)\ \text{(again)}$$

with horizontal strips containing one cell in column $4,2,1,1$ respectively. Then

$$\check{\mathfrak{h}}_{\{4\}}((1,4))=$$
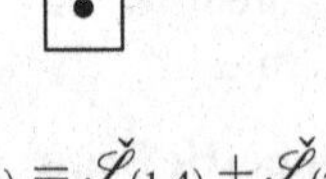
$$\check{\mathfrak{h}}_{\{2\}}((2,3))=$$
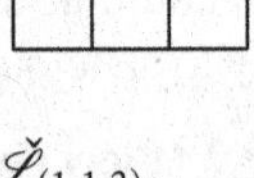
$$\check{\mathfrak{h}}_{\{1\}}((1,3,1))=$$
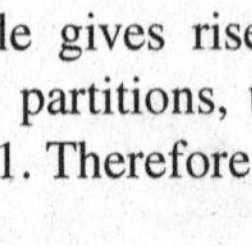
$$\check{\mathfrak{h}}_{\{1\}}((1,1,3))=$$
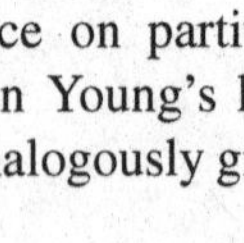

and hence

$$\check{\mathscr{S}}_{(1)}\check{\mathscr{S}}_{(1,3)}=\check{\mathscr{S}}_{(1,4)}+\check{\mathscr{S}}_{(2,3)}+\check{\mathscr{S}}_{(1,3,1)}+\check{\mathscr{S}}_{(1,1,3)}.$$

Classically, the Pieri rule gives rise to Young's lattice on partitions in the following way. Let λ,μ be partitions, then λ covers μ in Young's lattice if the coefficient of s_λ in $s_{(1)}s_\mu$ is 1. Therefore, Theorem 5.3.3 analogously gives rise to a poset on compositions.

Definition 5.3.5. Let α,β be compositions. Then β covers α in the poset $P_{\check{\mathscr{C}}}$ if the coefficient of $\check{\mathscr{S}}_\beta$ in $\check{\mathscr{S}}_{(1)}\check{\mathscr{S}}_\alpha$ is 1.

An analogue of the Littlewood-Richardson rule as stated in Theorem 3.2.13 also exists, but for this we need to define skew quasisymmetric Schur functions.

Definition 5.3.6. [15, Proposition 3.1] Let $D = \gamma /\!\!/_{\check{c}} \beta$ be a skew reverse composition shape. Then the *skew quasisymmetric Schur function* $\check{\mathscr{S}}_D$ is

$$\check{\mathscr{S}}_D = \sum_{\delta} \check{d}_{D\delta} F_\delta = \sum_{\check{\tau}} x^{\check{\tau}}$$

where the first sum is over all compositions $\delta \vDash |\gamma /\!\!/_{\check{c}} \beta|$ and $\check{d}_{D\delta}$ = the number of SRCTs $\check{\tau}$ of shape $\gamma /\!\!/_{\check{c}} \beta$ such that $\mathrm{Des}(\check{\tau}) = \mathrm{set}(\delta)$. The second sum is over all SSRCTs $\check{\tau}$ of shape $\gamma /\!\!/_{\check{c}} \beta$.

Example 5.3.7. We have $\check{\mathscr{S}}_{(1,2,3) /\!\!/_{\check{c}} (2)} = F_{(1,2,1)} + F_{(1,1,2)}$ from the following SRCTs.

1		
3	2	
•	•	4

1		
4	2	
•	•	3

The analogue of Theorem 3.2.13 can now be stated.

Theorem 5.3.8 (Littlewood-Richardson rule for quasisymmetric Schur functions). *[15, Theorem 3.5]Let γ, β be compositions. Then*

$$\check{\mathscr{S}}_{\gamma /\!\!/_{\check{c}} \beta} = \sum \check{C}^{\gamma}_{\alpha\beta} \check{\mathscr{S}}_\alpha$$

where the sum is over all compositions α, and $\check{C}^{\gamma}_{\alpha\beta}$ counts the number of SRCTs $\check{\tau}$ of shape $\gamma /\!\!/_{\check{c}} \beta$ such that using Schensted insertion for reverse tableaux

$$\check{\rho}_{\emptyset}^{-1}(\mathrm{P}(w_{col}(\check{\tau}))) = \check{U}_\alpha.$$

Corollary 5.3.9. *[15, Theorem 3.5] Let γ be a composition. Then*

$$\Delta \check{\mathscr{S}}_\gamma = \sum \check{C}^{\gamma}_{\alpha\beta} \check{\mathscr{S}}_\alpha \otimes \check{\mathscr{S}}_\beta$$

where the sum is over all compositions α, β and $\check{C}^{\gamma}_{\alpha\beta}$ counts the number of SRCTs $\check{\tau}$ of shape $\gamma /\!\!/_{\check{c}} \beta$ such that using Schensted insertion for reverse tableaux

$$\check{\rho}_{\emptyset}^{-1}(\mathrm{P}(w_{col}(\check{\tau}))) = \check{U}_\alpha.$$

Example 5.3.10. If we wish to compute $\check{\mathscr{S}}_{(2,4) /\!\!/_{\check{c}} (1)}$ we first compute all SRCTs of shape $(2,4) /\!\!/_{\check{c}} (1)$:

2	1		
•	5	4	3

3	2		
•	5	4	1

3	1		
•	5	4	2

4	1		
•	5	3	2

4	2		
•	5	3	1

with respective column reading words

$$21543 \quad 32541 \quad 31542 \quad 41532 \quad 42531$$

whose P-tableaux using Schensted insertion for reverse tableaux under $\check{\rho}_{\emptyset}^{-1}$ respectively gives

$$\begin{array}{|c|c|c} \hline 2 & 1 & \multicolumn{1}{c}{} \\ \hline 5 & 4 & \multicolumn{1}{c|}{3} \\ \hline \end{array} \quad \begin{array}{|c|c|c|} \hline 3 & 2 & 1 \\ \hline 5 & 4 & \multicolumn{1}{c}{} \\ \cline{1-2} \end{array} \quad \begin{array}{|c|c|c} \hline 3 & 1 & \multicolumn{1}{c}{} \\ \hline 5 & 4 & \multicolumn{1}{c|}{2} \\ \hline \end{array} \quad \begin{array}{|c|c|c|} \hline 4 & 3 & 2 \\ \hline 5 & 1 & \multicolumn{1}{c}{} \\ \cline{1-2} \end{array} \quad \begin{array}{|c|c|c|} \hline 4 & 3 & 1 \\ \hline 5 & 2 & \multicolumn{1}{c}{} \\ \cline{1-2} \end{array}$$

and hence $\check{\mathscr{S}}_{(2,4)/\!/_{\check{c}}(1)} = \check{\mathscr{S}}_{(2,3)} + \check{\mathscr{S}}_{(3,2)}$.

Note that a Littlewood-Richardson rule analogous to Theorem 3.2.16 would read quite differently since expanding the product of two generic quasisymmetric Schur functions in terms of quasisymmetric Schur functions often results in negative structure constants, for example,

$$\begin{aligned} \check{\mathscr{S}}_{(2,1)}\check{\mathscr{S}}_{(2,1)} = {} & \check{\mathscr{S}}_{(4,2)} + \check{\mathscr{S}}_{(4,1,1)} + 2\check{\mathscr{S}}_{(3,2,1)} + \check{\mathscr{S}}_{(3,1,2)} + 2\check{\mathscr{S}}_{(2,3,1)} \\ & + \check{\mathscr{S}}_{(1,3,2)} + \check{\mathscr{S}}_{(3,1,1,1)} + \check{\mathscr{S}}_{(2,2,2)} + \check{\mathscr{S}}_{(2,2,1,1)} + \check{\mathscr{S}}_{(2,1,2,1)} \\ & - \check{\mathscr{S}}_{(1,4,1)} - \check{\mathscr{S}}_{(1,3,1,1)} - \check{\mathscr{S}}_{(1,1,3,1)} - \check{\mathscr{S}}_{(1,2,2,1)}. \end{aligned}$$

5.4 Pieri and Littlewood-Richardson rules in QSym using Young composition tableaux

There exist Pieri rules for Young quasisymmetric Schur functions that can be stated analogously to the Pieri rules for Schur functions as given in Proposition 5.3.1. We need three new operators: $\hat{\mathfrak{d}}, \hat{\mathfrak{h}}, \hat{\mathfrak{v}}$. Informally, the decrementing $\hat{\mathfrak{d}}_s$ operator subtracts 1 from the leftmost part of size s in a composition, or returns the empty composition. Meanwhile the $\hat{\mathfrak{h}}_{\{s_1<\cdots<s_j\}}$ operator subtracts 1 from the leftmost part of size $s_j, s_{j-1}, \ldots$ recursively. Similarly, the $\hat{\mathfrak{v}}_{\{m_1\leqslant\cdots\leqslant m_j\}}$ operator subtracts 1 from the leftmost part of size $m_1, m_2, \ldots$ recursively.

Example 5.4.1. If $\alpha = (1,2,3)$, then

$$\hat{\mathfrak{h}}_{\{2,3\}}(\alpha) = \hat{\mathfrak{d}}_2(\hat{\mathfrak{d}}_3((1,2,3))) = \hat{\mathfrak{d}}_2((1,2,2)) = (1,1,2)$$

and

$$\hat{\mathfrak{v}}_{\{2,3\}}(\alpha) = \hat{\mathfrak{d}}_3(\hat{\mathfrak{d}}_2((1,2,3))) = \hat{\mathfrak{d}}_3((1,1,3)) = (1,1,2).$$

For completeness we define these three operators formally. Let $\alpha = (\alpha_1, \ldots, \alpha_k)$ be a composition whose largest part is m, and let $s \in [m]$. If there exists $1 \leqslant i \leqslant k$ such that $s = \alpha_i$ and $s \neq \alpha_j$ for all $j < i$, then define

$$\hat{\mathfrak{d}}_s(\alpha) = (\alpha_1, \ldots, \alpha_{i-1}, (s-1), \alpha_{i+1}, \ldots, \alpha_k),$$

otherwise define $\hat{\mathfrak{d}}_s(\alpha)$ to be the empty composition. Let $S = \{s_1 < \cdots < s_j\}$. Then define

$$\hat{\mathfrak{h}}_S(\alpha) = \hat{\mathfrak{d}}_{s_1}(\ldots(\hat{\mathfrak{d}}_{s_{j-1}}(\hat{\mathfrak{d}}_{s_j}(\alpha)))\ldots).$$

Similarly let $M = \{m_1 \leqslant \cdots \leqslant m_j\}$. Then define

$$\hat{\mathfrak{v}}_M(\alpha) = \hat{\mathfrak{d}}_{m_j}(\ldots(\hat{\mathfrak{d}}_{m_2}(\hat{\mathfrak{d}}_{m_1}(\alpha)))\ldots).$$

We remove any zeros from $\hat{\mathfrak{h}}_S(\alpha)$ or $\hat{\mathfrak{v}}_M(\alpha)$ to obtain a composition if needs be.

For any horizontal strip δ we denote by $S(\delta)$ the set of columns its skew diagram occupies, and for any vertical strip ε we denote by $M(\varepsilon)$ the multiset of columns its skew diagram occupies, where multiplicities for a column are given by the number of cells in that column, and column indices are listed in weakly increasing order. We are now ready to state our analogous Pieri rules.

Theorem 5.4.2 (Pieri rules for Young quasisymmetric Schur functions). *Let α be a composition. Then*

$$\hat{\mathscr{S}}_{(n)}\hat{\mathscr{S}}_{\alpha} = \sum_{\beta} \hat{\mathscr{S}}_{\beta}$$

where the sum is taken over all compositions β such that

1. *$\delta = \widetilde{\beta}/\widetilde{\alpha}$ is a horizontal strip,*
2. *$|\delta| = n$,*
3. *$\hat{\mathfrak{h}}_{S(\delta)}(\beta) = \alpha$.*

Also,

$$\hat{\mathscr{S}}_{(1^n)}\hat{\mathscr{S}}_{\alpha} = \sum_{\beta} \hat{\mathscr{S}}_{\beta}$$

where the sum is taken over all compositions β such that

1. *$\varepsilon = \widetilde{\beta}/\widetilde{\alpha}$ is a vertical strip,*
2. *$|\varepsilon| = n$,*
3. *$\hat{\mathfrak{v}}_{M(\varepsilon)}(\beta) = \alpha$.*

Proof. This follows immediately from Theorem 5.3.3 since the automorphism ρ : QSym $\to$ QSym where $\rho(\check{\mathscr{S}}_{\alpha^r}) = \hat{\mathscr{S}}_{\alpha}$ reverses the indexing compositions. □

We use Young composition diagrams in place of compositions in the next example to illustrate Theorem 5.4.2. Then $\hat{\mathfrak{d}}_s$ is the operation that removes the rightmost cell from the row of length s with smallest index.

Example 5.4.3. If we place • in the cell to be removed, then

$\hat{\mathfrak{d}}_1((3,1,1)) =$ 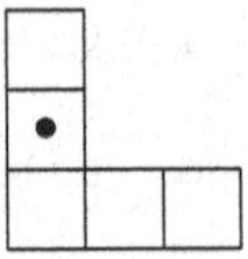 $= (3,1).$

If we wish to compute $\hat{\mathscr{S}}_{(1)}\hat{\mathscr{S}}_{(3,1)}$, then we consider the four skew diagrams

$$(4,1)/(3,1),\ (3,2)/(3,1),\ (3,1,1)/(3,1),\ (3,1,1)/(3,1)\ \text{(again)}$$

with horizontal strips containing one cell in column $4,2,1,1$ respectively. Then

$\hat{\mathfrak{h}}_{\{4\}}((4,1)) =$ 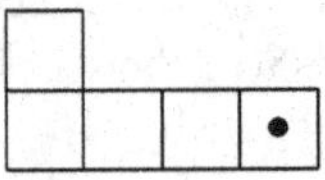 $\hat{\mathfrak{h}}_{\{2\}}((3,2)) =$ 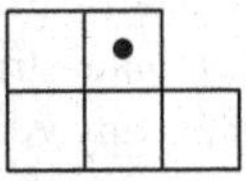

$\hat{\mathfrak{h}}_{\{1\}}((1,3,1)) =$ 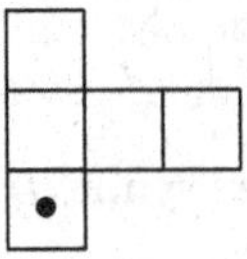 $\hat{\mathfrak{h}}_{\{1\}}((3,1,1)) =$ 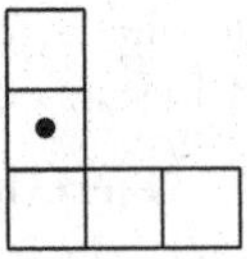

and hence

$$\hat{\mathscr{S}}_{(1)}\hat{\mathscr{S}}_{(3,1)} = \hat{\mathscr{S}}_{(4,1)} + \hat{\mathscr{S}}_{(3,2)} + \hat{\mathscr{S}}_{(1,3,1)} + \hat{\mathscr{S}}_{(3,1,1)}.$$

Note this is Example 5.3.4 with the indexing compositions reversed as described in the proof of Theorem 5.4.2.

As discussed in the previous section, Young's lattice can be seen to arise from the Pieri rules for symmetric functions: Let λ, μ be partitions, then λ covers μ in Young's lattice if the coefficient of s_λ in $s_{(1)}s_\mu$ is 1. Therefore, Theorem 5.4.2 analogously gives rise to a poset on compositions.

Definition 5.4.4. Let α, β be compositions. Then β covers α in the poset $P_{\hat{\mathcal{C}}}$ if the coefficient of $\hat{\mathscr{S}}_\beta$ in $\hat{\mathscr{S}}_{(1)}\hat{\mathscr{S}}_\alpha$ is 1.

We now define Young skew quasisymmetric Schur functions to be

$$\hat{\mathscr{S}}_{\gamma /\!/_{\hat{\mathcal{C}}} \beta} = \rho(\check{\mathscr{S}}_{\gamma^r /\!/_{\check{\mathcal{C}}} \beta^r}) \tag{5.3}$$

in order to describe an analogue of the Littlewood-Richardson rule given in Theorem 3.2.13, but first we give a combinatorial description of them.

Proposition 5.4.5. *Let $D = \gamma /\!/_{\hat{\mathcal{C}}} \beta$ be a skew Young composition shape. Then the* skew Young quasisymmetric Schur function $\hat{\mathscr{S}}_D$ *is*

$$\hat{\mathscr{S}}_D = \sum_\delta \hat{d}_{D\delta} F_\delta = \sum_\tau x^\tau$$

where the first sum is over all compositions $\delta \models |\gamma /\!\!/_{\hat{c}} \beta|$ *and* $\hat{d}_{D\delta} =$ *the number of SYCTs* τ *of shape* $\gamma /\!\!/_{\hat{c}} \beta$ *such that* $\mathrm{Des}(\tau) = \mathrm{set}(\delta)$*. The second sum is over all SSYCTs* τ *of shape* $\gamma /\!\!/_{\hat{c}} \beta$*.*

Proof. From Equation (5.3) and Definition 5.3.6

$$\hat{\mathscr{S}}_{\gamma /\!\!/_{\hat{c}} \beta} = \rho(\check{\mathscr{S}}_{\gamma^r /\!\!/_{\check{c}} \beta^r}) = \rho\left(\sum_{\delta} \check{d}_{\gamma^r /\!\!/_{\check{c}} \beta^r \delta^r} F_{\delta^r}\right) = \sum_{\delta} \check{d}_{\gamma^r /\!\!/_{\check{c}} \beta^r \delta^r} F_{\delta}$$

where

$$\check{d}_{\gamma^r /\!\!/_{\check{c}} \beta^r \delta^r} = \#\{\check{\tau} \in SRCTs \mid sh(\check{\tau}) = \gamma^r /\!\!/_{\check{c}} \beta^r, \mathrm{Des}(\check{\tau}) = \mathrm{set}(\delta^r)\}.$$

However,

$$\check{d}_{\gamma^r /\!\!/_{\check{c}} \beta^r \delta^r} = \#\{\tau \in SYCTs \mid sh(\tau) = \gamma /\!\!/_{\hat{c}} \beta, \mathrm{Des}(\tau) = \mathrm{set}(\delta)\} = \hat{d}_{\gamma /\!\!/_{\hat{c}} \beta \delta}$$

since by Proposition 4.3.5 the bijection $\hat{\Gamma} : SYCTs \rightarrow SRCTs$ reverses both indexing shape and descent set. For the second equation the proofs of Propositions 5.2.4 and 5.2.6 can be used with skew composition shapes in place of straight shapes. □

Example 5.4.6. By Equation (5.3) and Example 5.3.7 we have

$$\hat{\mathscr{S}}_{(3,2,1) /\!\!/_{\hat{c}} (2)} = \rho(\check{\mathscr{S}}_{(1,2,3) /\!\!/_{\check{c}} (2)}) = \rho(F_{(1,2,1)} + F_{(1,1,2)}) = F_{(1,2,1)} + F_{(2,1,1)}$$

or we can compute it directly from the following SYCTs.

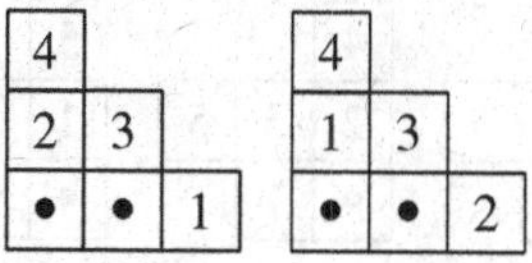

The interested reader may wish to compare the computation of $\hat{C}^{\gamma}_{\alpha\beta}$ below with the computation of $\check{C}^{\gamma}_{\alpha\beta}$.

Theorem 5.4.7 (Littlewood-Richardson rule for Young quasisymmetric Schur functions). *Let* γ, α *be compositions. Then*

$$\hat{\mathscr{S}}_{\gamma /\!\!/_{\hat{c}} \alpha} = \sum \hat{C}^{\gamma}_{\alpha\beta} \hat{\mathscr{S}}_{\beta}$$

where the sum is over all compositions β*, and* $\hat{C}^{\gamma}_{\alpha\beta}$ *counts the number of SYCTs* τ *of shape* $\gamma /\!\!/_{\hat{c}} \alpha$ *such that using Schensted insertion*

$$\hat{\rho}_{\emptyset}^{-1}(\mathrm{P}(w_{col}(\tau))) = U_{\beta}.$$

Proof. By Equation (5.3) and Theorem 5.3.8 we have

$$\hat{\mathscr{S}}_{\gamma /\!/_{\hat{c}} \alpha} = \rho(\check{\mathscr{S}}_{\gamma^r /\!/_{\check{c}} \alpha^r}) = \rho\left(\sum_{\beta} \check{C}^{\gamma^r}_{\beta^r \alpha^r} \check{\mathscr{S}}_{\beta^r}\right) = \sum_{\beta} \check{C}^{\gamma^r}_{\beta^r \alpha^r} \hat{\mathscr{S}}_{\beta}.$$

However, $\check{C}^{\gamma^r}_{\beta^r \alpha^r}$ is the number of SRCTs of shape $\gamma^r /\!/_{\check{c}} \alpha^r$ whose P-tableau using Schensted insertion for reverse tableaux under $\check{\rho}_{\emptyset}^{-1}$ yields $\check{U}_{\beta^r}$. Due to reversal of indices by $\hat{\Gamma}$, $\hat{\Gamma}^{-1}(\check{\tau})$ of such an SRCT $\check{\tau}$ will be an SYCT of shape $\gamma /\!/_{\hat{c}} \alpha$ whose P-tableau using Schensted insertion for Young tableaux under $\hat{\rho}_{\emptyset}^{-1}$ yields $U_{\beta} = \hat{\Gamma}^{-1}(\check{U}_{\beta^r})$. That is,

$$\check{C}^{\gamma^r}_{\beta^r \alpha^r} = \hat{C}^{\gamma}_{\alpha\beta}.$$

□

Corollary 5.4.8. *Let γ be a composition. Then*

$$\Delta \hat{\mathscr{S}}_{\gamma} = \sum \hat{C}^{\gamma}_{\alpha\beta} \hat{\mathscr{S}}_{\beta} \otimes \hat{\mathscr{S}}_{\alpha}$$

where the sum is over all compositions α, β and $\hat{C}^{\gamma}_{\alpha\beta}$ counts the number of SYCTs τ of shape $\gamma /\!/_{\hat{c}} \alpha$ such that using Schensted insertion

$$\hat{\rho}_{\emptyset}^{-1}(\mathrm{P}(w_{col}(\tau))) = U_{\beta}.$$

Example 5.4.9. If we wish to compute $\hat{\mathscr{S}}_{(2,3,1)/\!/_{\hat{c}}(1)}$ we first compute all SYCTs of shape $\hat{\mathscr{S}}_{(2,3,1)/\!/_{\hat{c}}(1)}$:

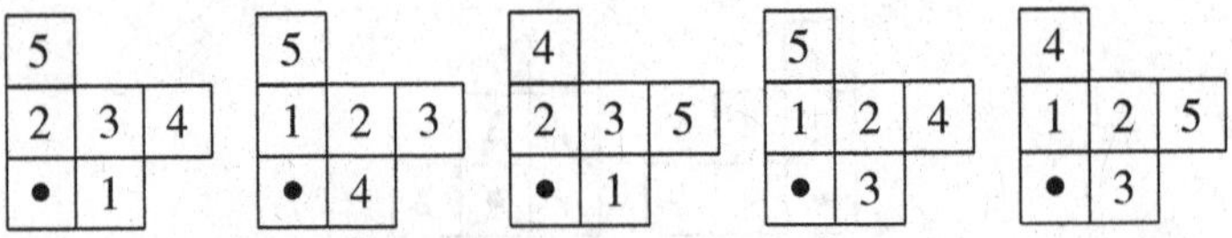

with respective column reading words

52314 51423 42315 51324 41325

whose P-tableau using Schensted insertion for Young tableaux under $\hat{\rho}_{\emptyset}^{-1}$ respectively gives

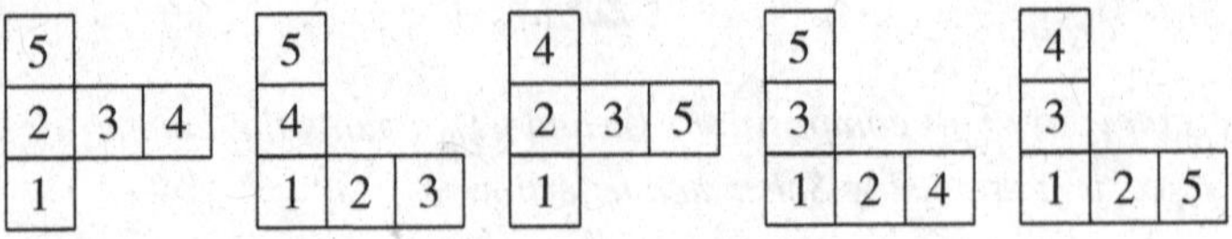

and hence $\hat{\mathscr{S}}_{(2,3,1)/\!/_{\hat{c}}(1)} = \hat{\mathscr{S}}_{(1,3,1)} + \hat{\mathscr{S}}_{(3,1,1)}$.

A Littlewood-Richardson rule analogous to Theorem 3.2.16 would be different in flavour, since expanding the product of two generic quasisymmetric Schur functions in terms of quasisymmetric Schur functions often results in negative structure constants, as can be seen by applying ρ to the last equation in the previous section.

5.5 Pieri and Littlewood-Richardson rules in NSym using reverse composition tableaux

Using the pairing of dual bases given is Subsection 3.4.2, namely,

$$\langle F_\alpha, \mathbf{r}_\beta \rangle = \delta_{\alpha\beta}$$

we can define noncommutative Schur functions.

Definition 5.5.1. Let α be a composition. Then the *noncommutative Schur function* $\check{\mathbf{s}}_\alpha$ is defined by

$$\langle \check{\mathscr{S}}_\beta, \check{\mathbf{s}}_\alpha \rangle = \delta_{\beta\alpha}$$

for every composition β.

The basis of quasisymmetric Schur functions $\{\check{\mathscr{S}}_\alpha\}_{\alpha\vDash n\geqslant 0}$ of QSym is dual to the basis of noncommutative Schur functions $\{\check{\mathbf{s}}_\alpha\}_{\alpha\vDash n\geqslant 0}$ of NSym by construction, and moreover, [15, Equation (2.12)]

$$\chi(\check{\mathbf{s}}_\alpha) = s_{\widetilde{\alpha}} \tag{5.4}$$

and [15, Equation (3.6)]

$$\langle \check{\mathscr{S}}_{\gamma /\!/_{\check{c}} \beta}, \check{\mathbf{s}}_\alpha \rangle = \langle \check{\mathscr{S}}_\gamma, \check{\mathbf{s}}_\alpha \check{\mathbf{s}}_\beta \rangle = \check{C}^{\gamma}_{\alpha\beta}. \tag{5.5}$$

From this latter equation we immediately get the following theorem and corollaries.

Theorem 5.5.2 (Littlewood-Richardson rule for noncommutative Schur functions). *[15, Theorem 3.5]Let α, β be compositions. Then*

$$\check{\mathbf{s}}_\alpha \check{\mathbf{s}}_\beta = \sum \check{C}^{\gamma}_{\alpha\beta} \check{\mathbf{s}}_\gamma$$

where the sum is over all compositions γ, and $\check{C}^{\gamma}_{\alpha\beta}$ counts the number of SRCTs $\check{\tau}$ of shape $\gamma /\!/_{\check{c}} \beta$ such that using reverse Schensted insertion

$$\check{\rho}_\emptyset^{-1}(\mathrm{P}(w_{col}(\check{\tau}))) = \check{U}_\alpha.$$

Corollary 5.5.3. *[15, Corollary 3.7] Let α and β be compositions with $\lambda = \widetilde{\alpha}$ and $\mu = \widetilde{\beta}$, and let ν be a partition. Then $\chi(\check{\mathbf{s}}_\alpha \check{\mathbf{s}}_\beta) = s_\lambda s_\mu$, and*

$$c^{\nu}_{\lambda\mu} = \sum_{\widetilde{\gamma}=\nu} \check{C}^{\check{\gamma}}_{\alpha\beta}. \tag{5.6}$$

Corollary 5.5.4 (Pieri rules for noncommutative Schur functions). *[15, Corollary 3.8] Let β be a composition. Then*

$$\check{\mathbf{s}}_{(n)}\check{\mathbf{s}}_{\beta} = \sum_{\gamma} \check{\mathbf{s}}_{\gamma}$$

where γ runs over all compositions satisfying

$$\beta = \gamma^0 \lessdot_{\check{c}} \gamma^1 \cdots \lessdot_{\check{c}} \gamma^n = \gamma$$

and

$$\mathrm{col}(\gamma^0 \lessdot_{\check{c}} \gamma^1 \cdots \lessdot_{\check{c}} \gamma^n)$$

is strictly increasing.

Let β be a composition. Then

$$\check{\mathbf{s}}_{(1^n)}\check{\mathbf{s}}_{\beta} = \sum_{\gamma} \check{\mathbf{s}}_{\gamma}$$

where γ runs over all compositions satisfying

$$\beta = \gamma^0 \lessdot_{\check{c}} \gamma^1 \cdots \lessdot_{\check{c}} \gamma^n = \gamma$$

and

$$\mathrm{col}(\gamma^0 \lessdot_{\check{c}} \gamma^1 \cdots \lessdot_{\check{c}} \gamma^n)$$

is weakly decreasing.

Pictorially, in the first case we can think of γ as being obtained from β by adding n cells from left to right using the cover relations of $\mathcal{L}_{\check{c}}$, with no two cells in the same column. In the second case we can think of γ as being obtained from β by adding n cells from right to left using the cover relations of $\mathcal{L}_{\check{c}}$, with no two cells in the same row.

Example 5.5.5. For the Littlewood-Richardson rule we compute

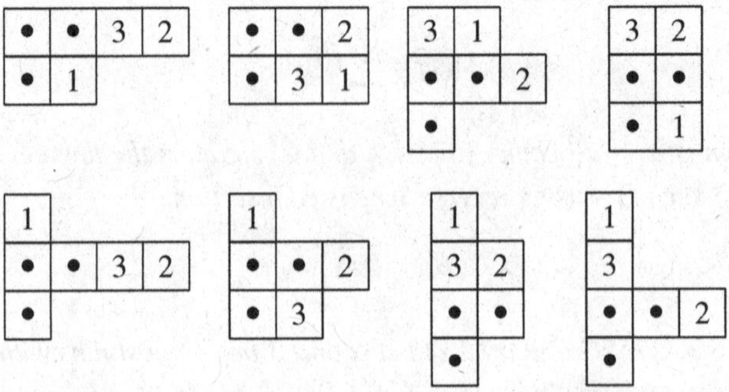

$$\check{\mathbf{s}}_{(1,2)}\check{\mathbf{s}}_{(2,1)} = \check{\mathbf{s}}_{(4,2)} + \check{\mathbf{s}}_{(3,3)} + \check{\mathbf{s}}_{(2,3,1)} + \check{\mathbf{s}}_{(2,2,2)} + \check{\mathbf{s}}_{(1,4,1)} + \check{\mathbf{s}}_{(1,3,2)} + \check{\mathbf{s}}_{(1,2,2,1)} + \check{\mathbf{s}}_{(1,1,3,1)}$$

while the Pieri rule is a simpler computation.

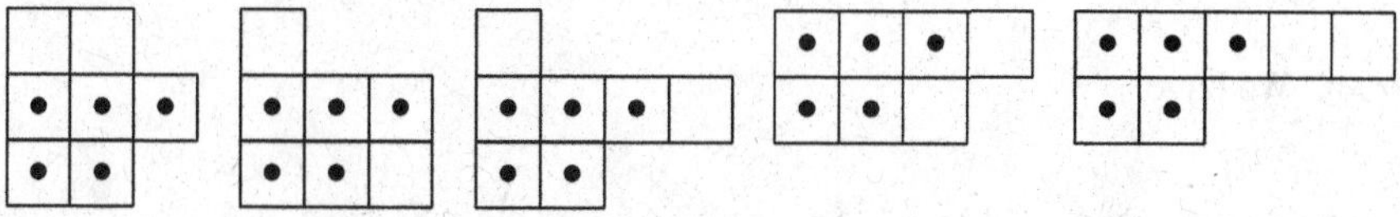

$$\check{\mathbf{s}}_{(2)}\check{\mathbf{s}}_{(3,2)} = \check{\mathbf{s}}_{(2,3,2)} + \check{\mathbf{s}}_{(1,3,3)} + \check{\mathbf{s}}_{(1,4,2)} + \check{\mathbf{s}}_{(4,3)} + \check{\mathbf{s}}_{(5,2)}$$

5.6 Pieri and Littlewood-Richardson rules in NSym using Young composition tableaux

We can define Young noncommutative Schur functions using the pairing of dual bases from Subsection 3.4.2, that is,

$$\langle F_\alpha, \mathbf{r}_\beta \rangle = \delta_{\alpha\beta}.$$

Definition 5.6.1. Let α be a composition. Then the *Young noncommutative Schur function* $\hat{\mathbf{s}}_\alpha$ is defined by

$$\langle \hat{\mathscr{S}}_\beta, \hat{\mathbf{s}}_\alpha \rangle = \delta_{\beta\alpha}$$

for every composition β.

The basis of Young quasisymmetric Schur functions $\{\hat{\mathscr{S}}_\alpha\}_{\alpha\vDash n\geqslant 0}$ of QSym is dual to the basis of Young noncommutative Schur functions $\{\hat{\mathbf{s}}_\alpha\}_{\alpha\vDash n\geqslant 0}$ of NSym by construction. Additionally, we have

$$\hat{\mathbf{s}}_\alpha = \rho(\check{\mathbf{s}}_{\alpha^r}) \tag{5.7}$$

and thus by Equation (5.4) and Equation (3.29)

$$\chi(\hat{\mathbf{s}}_\alpha) = s_{\widetilde{\alpha}} \tag{5.8}$$

and

$$\langle \hat{\mathscr{S}}_{\gamma /\!/_{\hat{c}} \alpha}, \hat{\mathbf{s}}_\beta \rangle = \langle \hat{\mathscr{S}}_\gamma, \hat{\mathbf{s}}_\alpha \hat{\mathbf{s}}_\beta \rangle = \hat{C}^{\gamma}_{\alpha\beta}. \tag{5.9}$$

This leads to the following theorem and corollaries.

Theorem 5.6.2 (Littlewood-Richardson rule for Young noncommutative Schur functions). *Let α, β be compositions. Then*

$$\hat{\mathbf{s}}_\alpha \hat{\mathbf{s}}_\beta = \sum \hat{C}^{\gamma}_{\alpha\beta} \hat{\mathbf{s}}_\gamma$$

where the sum is over all compositions γ, and $\hat{C}^{\gamma}_{\alpha\beta}$ counts the number of SYCTs τ of shape $\gamma /\!/_{\hat{c}} \alpha$ such that using Schensted insertion

$$\hat{\rho}_{\emptyset}^{-1}(\mathrm{P}(w_{col}(\tau))) = U_{\beta}.$$

Proof. We have

$$\hat{\mathbf{s}}_{\alpha}\hat{\mathbf{s}}_{\beta} = \rho(\check{\mathbf{s}}_{\beta^r}\check{\mathbf{s}}_{\alpha^r}) = \rho\left(\sum_{\gamma} \check{C}^{\gamma^r}_{\beta^r\alpha^r}\check{\mathbf{s}}_{\gamma^r}\right) = \sum_{\gamma} \hat{C}^{\gamma}_{\alpha\beta}\hat{\mathbf{s}}_{\gamma}$$

by Theorem 5.5.2 and the proof of Theorem 5.4.7. □

Corollary 5.6.3. *Let α and β be compositions with $\lambda = \widetilde{\alpha}$ and $\mu = \widetilde{\beta}$, and let ν be a partition. Then $\chi(\hat{\mathbf{s}}_{\alpha}\hat{\mathbf{s}}_{\beta}) = s_{\lambda}s_{\mu}$, and*

$$c^{\nu}_{\lambda\mu} = \sum_{\widetilde{\gamma}=\nu} \hat{C}^{\gamma}_{\alpha\beta}. \tag{5.10}$$

Corollary 5.6.4 (Pieri rules for Young noncommutative Schur functions). *Let α be a composition. Then*

$$\hat{\mathbf{s}}_{\alpha}\hat{\mathbf{s}}_{(n)} = \sum_{\gamma} \hat{\mathbf{s}}_{\gamma}$$

where γ runs over all compositions satisfying

$$\alpha = \gamma^0 \lessdot_{\hat{c}} \gamma^1 \cdots \lessdot_{\hat{c}} \gamma^n = \gamma$$

and

$$\mathrm{col}(\gamma^0 \lessdot_{\hat{c}} \gamma^1 \cdots \lessdot_{\hat{c}} \gamma^n)$$

is strictly increasing.

Let α be a composition. Then

$$\hat{\mathbf{s}}_{\alpha}\hat{\mathbf{s}}_{(1^n)} = \sum_{\gamma} \hat{\mathbf{s}}_{\gamma}$$

where γ runs over all compositions satisfying

$$\alpha = \gamma^0 \lessdot_{\hat{c}} \gamma^1 \cdots \lessdot_{\hat{c}} \gamma^n = \gamma$$

and

$$\mathrm{col}(\gamma^0 \lessdot_{\hat{c}} \gamma^1 \cdots \lessdot_{\hat{c}} \gamma^n)$$

is weakly decreasing.

Pictorially, in the first case we can think of γ as being obtained from α by adding n cells from left to right using the cover relations of $\mathscr{L}_{\hat{c}}$, with no two cells in the same column. In the second case we can think of γ as being obtained from α by adding n cells from right to left using the cover relations of $\mathscr{L}_{\hat{c}}$, with no two cells in the same row.

Example 5.6.5. For the Littlewood-Richardson rule we compute

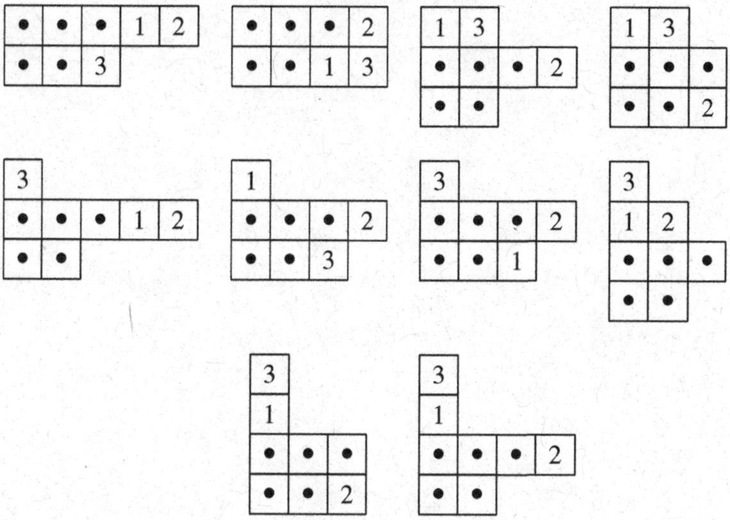

$$\hat{\mathbf{s}}_{(2,3)}\hat{\mathbf{s}}_{(2,1)} = \hat{\mathbf{s}}_{(3,5)} + \hat{\mathbf{s}}_{(4,4)} + \hat{\mathbf{s}}_{(2,4,2)} + \hat{\mathbf{s}}_{(3,3,2)} + \hat{\mathbf{s}}_{(2,5,1)} + 2\hat{\mathbf{s}}_{(3,4,1)} + \hat{\mathbf{s}}_{(2,3,2,1)} + \hat{\mathbf{s}}_{(3,3,1,1)} + \hat{\mathbf{s}}_{(2,4,1,1)}$$

while the Pieri rule is a simpler computation.

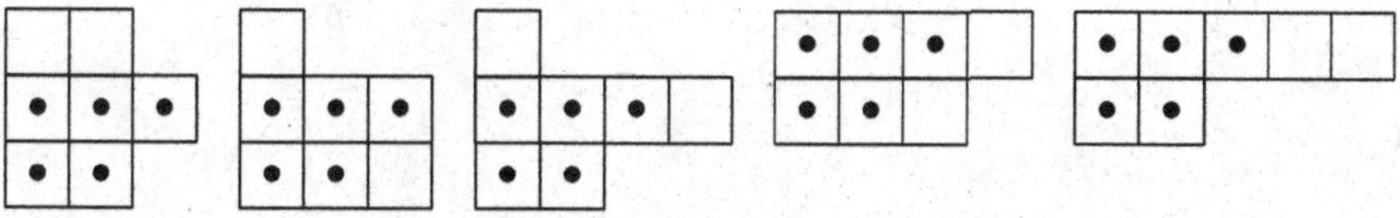

$$\hat{\mathbf{s}}_{(2,3)}\hat{\mathbf{s}}_{(2)} = \hat{\mathbf{s}}_{(2,3,2)} + \hat{\mathbf{s}}_{(3,3,1)} + \hat{\mathbf{s}}_{(2,4,1)} + \hat{\mathbf{s}}_{(3,4)} + \hat{\mathbf{s}}_{(2,5)}$$

References

1. Aguiar, M.: Infinitesimal Hopf algebras and the **cd**-index of polytopes. Discrete Comput. Geom. **27**, 3–28 (2002)
2. Aguiar, M., Hsiao, S.: Canonical characters on quasi-symmetric functions and bivariate Catalan numbers. Electron. J. Combin. **11**, Paper 15, 34 pp (2004/06)
3. Aguiar, M., Sottile, F.: Structure of the Malvenuto-Reutenauer Hopf algebra of permutations. Adv. Math. **191**, 225–275 (2005)
4. Aguiar, M., Bergeron, N., Sottile, F.: Combinatorial Hopf algebras and generalized Dehn-Sommerville relations. Compos. Math. **142**, 1–30 (2006)
5. Assaf, S.: On dual equivalence and Schur positivity. Preprint. arXiv:1107.0090v1[math.CO]
6. Aval, J.-C., Bergeron, F., Bergeron, N.: Ideals of quasi-symmetric functions and super-covariant polynomials for S_n. Adv. Math. **181**, 353–367 (2004)
7. Baker, A., Richter, B.: Quasisymmetric functions from a topological point of view. Math. Scand. **103**, 208–242 (2008)
8. Baumann, P., Hohlweg, C.: A Solomon descent theory for the wreath products $G \wr \mathfrak{S}_n$. Trans. Amer. Math. Soc. **360**, 1475–1538 (2008)
9. Bender, E., Knuth, D.: Enumeration of plane partitions. J. Combin. Theory Ser. A **13**, 40–54 (1972)
10. Bergeron, N., Zabrocki, M.: The Hopf algebras of symmetric functions and quasi-symmetric functions in non-commutative variables are free and co-free. J. Algebra Appl. **8**, 581–600 (2009)
11. Bergeron, N., Reutenauer, C., Rosas, M., Zabrocki, M.: Invariants and coinvariants of the symmetric groups in noncommuting variables. Canad. J. Math. **60**, 266–296 (2008)
12. Bergeron, N., Mykytiuk, S., Sottile, F., van Willigenburg, S.: Noncommutative Pieri operators on posets. J. Combin. Theory Ser. A **91**, 84–110 (2000)
13. Bergeron, N., Mykytiuk, S., Sottile, F., van Willigenburg, S.: Shifted quasi-symmetric functions and the Hopf algebra of peak functions. Discrete Math. **246**, 57–66 (2002)
14. Bessenrodt, C., van Willigenburg, S.: Multiplicity free Schur, skew Schur, and quasisymmetric Schur functions. Preprint, to appear Ann. Comb. arXiv:1105.4212v2[math.CO]
15. Bessenrodt, C., Luoto, K., van Willigenburg, S.: Skew quasisymmetric Schur functions and noncommutative Schur functions. Adv. Math. **226**, 4492–4532 (2011)
16. Billera, L., Brenti, F.: Quasisymmetric functions and Kazhdan-Lusztig polynomials. Israel J. Math. **184**, 317–348 (2011)
17. Billera, L., Hsiao, S., van Willigenburg, S.: Peak quasisymmetric functions and Eulerian enumeration. Adv. Math. **176**, 248–276 (2003)
18. Billera, L., Thomas, H., van Willigenburg, S.: Decomposable compositions, symmetric quasisymmetric functions and equality of ribbon Schur functions. Adv. Math. **204**, 204–240 (2006)

K. Luoto et al., *An Introduction to Quasisymmetric Schur Functions*,
SpringerBriefs in Mathematics, DOI 10.1007/978-1-4614-7300-8,

© Kurt Luoto, Stefan Mykytiuk, Stephanie van Willigenburg 2013

19. Billera, L., Jia, N., Reiner, V.: A quasisymmetric function for matroids. European J. Combin. **30**, 1727–1757 (2009)
20. Billey, S., Haiman, M.: Schubert polynomials for the classical groups. J. Amer. Math. Soc. **8**, 443–482 (1995)
21. Blessenohl, D., Schocker, M.: Noncommutative character theory of the symmetric group. Imperial College Press, London (2005)
22. Buchstaber,V., Erokhovets, N.: Polytopes, Hopf algebras and Quasi-symmetric functions. Preprint. arXiv:1011.1536v1[math.CO]
23. Chow, T.: Descents, quasi-symmetric functions, Robinson-Schensted for posets, and the chromatic symmetric function. J. Algebraic Combin. **10**, 227–240 (1999)
24. Derksen, H.: Symmetric and quasi-symmetric functions associated to polymatroids. J. Algebraic Combin. **30**, 43–86 (2009)
25. Dimakis, A., Müller-Hoissen, F.: Quasi-symmetric functions and the KP hierarchy. J. Pure Appl. Algebra **214**, 449–460 (2010)
26. Duchamp, G., Hivert, F., Thibon, J.-Y.: Noncommutative symmetric functions. VI. Free quasi-symmetric functions and related algebras. Internat. J. Algebra Comput. **12**, 671–717 (2002)
27. Ehrenborg, R.: On posets and Hopf algebras. Adv. Math. **119**, 1–25 (1996)
28. Ehrenborg, R., Readdy, M.: The Tchebyshev transforms of the first and second kind. Ann. Comb. **14**, 211–244 (2010)
29. Ferreira, J.: A Littlewood-Richardson Type Rule for Row-Strict Quasisymmetric Schur Functions. Preprint. arXiv:1102.1458v1[math.CO]
30. Foissy, L.: Faà di Bruno subalgebras of the Hopf algebra of planar trees from combinatorial Dyson-Schwinger equations. Adv. Math. **218**, 136–162 (2008)
31. Fulton, W.: Young tableaux. With applications to representation theory and geometry. Cambridge University Press, Cambridge (1997)
32. Garsia, A., Reutenauer, C.: A decomposition of Solomon's descent algebra. Adv. Math. **77**, 189–262 (1989)
33. Garsia, A., Wallach, N.: $r-$QSym is free over Sym. J. Combin. Theory Ser. A **114**, 704–732 (2007)
34. Gelfand, I., Krob, D., Lascoux, A., Leclerc, B., Retakh, V., Thibon, J.-Y.: Noncommutative symmetric functions. Adv. Math. **112**, 218–348 (1995)
35. Gessel, I.: Multipartite *P*-partitions and inner products of skew Schur functions. Contemp. Math. **34**, 289–301 (1984)
36. Gessel, I., Reutenauer, C.: Counting permutations with given cycle structure and descent set. J. Combin. Theory Ser. A **64**, 189–215 (1993)
37. Gnedin, A., Olshanski, G.: Coherent permutations with descent statistic and the boundary problem for the graph of zigzag diagrams. Int. Math. Res. Not. Art. ID 51968, 39 pp (2006)
38. Haglund, J., Haiman, M., Loehr, N.: A combinatorial formula for Macdonald polynomials. J. Amer. Math. Soc. **18**, 735–761 (2005)
39. Haglund, J., Luoto, K., Mason, S., van Willigenburg, S.: Refinements of the Littlewood-Richardson rule. Trans. Amer. Math. Soc. **363**, 1665–1686 (2011)
40. Haglund, J., Luoto, K., Mason, S., van Willigenburg, S.: Quasisymmetric Schur functions. J. Combin. Theory Ser. A **118**, 463–490 (2011)
41. Hazewinkel, M.: The Leibniz-Hopf algebra and Lyndon words. CWI Report Dept. of Analysis, Algebra and Geometry. AM-R9612 (1996)
42. Hazewinkel, M.: Symmetric functions, noncommutative symmetric functions, and quasisymmetric functions. Acta Appl. Math. **75**, 55–83 (2003)
43. Hazewinkel, M.: Symmetric functions, noncommutative symmetric functions and quasisymmetric functions. II. Acta Appl. Math. **85**, 319–340 (2005)
44. Hazewinkel, M.: Explicit polynomial generators for the ring of quasisymmetric functions over the integers. Acta Appl. Math. **109**, 39–44 (2010)
45. Hersh, P., Hsiao, S.: Random walks on quasisymmetric functions. Adv. Math. **222**, 782–808 (2009)

46. Hivert, F.: Hecke algebras, difference operators, and quasi-symmetric functions. Adv. Math. **155**, 181–238 (2000)
47. Hoffman, M.: (Non)commutative Hopf algebras of trees and (quasi)symmetric functions. In: Renormalization and Galois theories, pp. 209–227. IRMA Lect. Math. Theor. Phys., 15, Eur. Math. Soc., Zürich (2009)
48. Hsiao, S., Karaali, G.: Multigraded combinatorial Hopf algebras and refinements of odd and even subalgebras. J. Algebraic Combin. **34**, 451–506 (2011)
49. Hsiao, S., Petersen, (T.) K.: Colored posets and colored quasisymmetric functions. Ann. Comb. **14**, 251–289 (2010)
50. Humpert, B.: A quasisymmetric function generalization of the chromatic symmetric function. Electron. J. Combin. **18**, Paper 31, 13 pp (2011)
51. Krob, D., Leclerc, B., Thibon, J.-Y.: Noncommutative symmetric functions. II. Transformations of alphabets. Internat. J. Algebra Comput. **7**, 181–264 (1997)
52. Kwon, J.-H.: Crystal graphs for general linear Lie superalgebras and quasi-symmetric functions. J. Combin. Theory Ser. A **116**, 1199–1218 (2009)
53. Lam, T., Pylyavskyy, P.: *P*-partition products and fundamental quasi-symmetric function positivity. Adv. in Appl. Math. **40**, 271–294 (2008)
54. Lascoux, A., Novelli, J.-C., Thibon, J.-Y.: Noncommutative symmetric functions with matrix parameters. Preprint. arXiv:1110.3209v1[math.CO]
55. Lauve, A., Mason, S.: QSym over Sym has a stable basis. J. Combin. Theory Ser. A **118**, 1661–1673 (2011)
56. Littlewood, D., Richardson, A.: Group characters and algebra. Philos. Trans. R. Soc. Lond. Ser. A Math. Phys. Eng. Sci. **233**, 99–141 (1934)
57. Loehr, N., Warrington, G.: Quasisymmetric expansions of Schur-function plethysms. Proc. Amer. Math. Soc. **140**, 1159–1171 (2012)
58. Loehr, N., Serrano, L., Warrington, G.: Transition matrices for symmetric and quasisymmetric Hall-Littlewood polynomials. Preprint. arXiv:1202.3411v1[math.CO]
59. Luoto, K.: A matroid-friendly basis for the quasisymmetric functions. J. Combin. Theory Ser. A **115**, 777–798 (2008)
60. Macdonald, I.: Symmetric functions and Hall polynomials. Second edition. Oxford University Press, New York (1995)
61. MacMahon, P.: Combinatory analysis. Vol. I, II (bound in one volume). Reprint of An introduction to combinatory analysis (1920) and Combinatory analysis. Vol. I, II (1915, 1916). Dover Publications, Mineola (2004)
62. Malvenuto, C., Reutenauer, C.: Duality between quasi-symmetric functions and the Solomon descent algebra. J. Algebra **177**, 967–982 (1995)
63. Malvenuto, C., Reutenauer, C.: Plethysm and conjugation of quasi-symmetric functions. Discrete Math. **193**, 225–233 (1998)
64. Mason, S.: A decomposition of Schur functions and an analogue of the Robinson-Schensted-Knuth algorithm. Sém. Lothar. Combin. **57**, Art. B57e, 24 pp (2006/08)
65. Mason, S., Remmel, J.: Row-strict quasisymmetric Schur functions. Preprint. arXiv:1110.4014v1[math.CO]
66. McNamara, P.: EL-labelings, supersolvability and 0-Hecke algebra actions on posets. J. Combin. Theory Ser. A **101**, 69–89 (2003)
67. Menous, F., Novelli, J.-C., Thibon, J.-Y.: Mould calculus, polyhedral cones, and characters of combinatorial Hopf algebras. Preprint. arXiv:1109.1634v2[math.CO]
68. Milnor, J., Moore, J.: On the structure of Hopf algebras. Ann. of Math. **81**, 211–264 (1965)
69. Montgomery, S.: Hopf algebras and their actions on rings. American Mathematical Society, Providence (1993)
70. Novelli, J.-C., Schilling, A.: The forgotten monoid. In: Combinatorial representation theory and related topics, pp. 71–83. RIMS Kokyuroku Bessatsu B8, Kyoto (2008)
71. Reutenauer, C.: Free Lie algebras. Oxford University Press, New York (1993)
72. Sagan, B.: The symmetric group. Representations, combinatorial algorithms, and symmetric functions. Second edition. Springer-Verlag, New York (2001)

73. Schützenberger, M.-P.: La correspondance de Robinson. In: Combinatoire et représentation du groupe symétrique (Actes Table Ronde CNRS, Univ. Louis-Pasteur Strasbourg, Strasbourg, 1976) Lecture Notes in Math., Vol. 579, pp. 59–113. Springer, Berlin (1977)
74. Shareshian, J., Wachs, M.: Eulerian quasisymmetric functions. Adv. Math. **225**, 2921–2966 (2010)
75. Shareshian, J., Wachs, M.: Chromatic quasisymmetric functions and Hessenberg varieties. Preprint. arXiv:1106.4287v3[math.CO]
76. Solomon, L.: A Mackey formula in the group ring of a Coxeter group. J. Algebra **41**, 255–264 (1976)
77. Stanley, R.: Ordered structures and partitions. Mem. Amer. Math. Soc. **119**, (1972)
78. Stanley, R.: On the number of reduced decompositions of elements of Coxeter groups. European J. Combin. **5**, 359–372 (1984)
79. Stanley, R.: A symmetric function generalization of the chromatic polynomial of a graph. Adv. Math. **111**, 166–194 (1995)
80. Stanley, R.: Flag-symmetric and locally rank-symmetric partially ordered sets. Electron. J. Combin. **3**, Paper 6, 22 pp (1996)
81. Stanley, R.: Enumerative combinatorics. Vol. 2. Cambridge University Press, Cambridge (1999)
82. Stanley, R.: Generalized riffle shuffles and quasisymmetric functions. Ann. Comb. **5**, 479–491 (2001)
83. Stembridge, J.: Enriched *P*-partitions. Trans. Amer. Math. Soc. **349**, 763–788 (1997)
84. Sweedler, M.: Hopf algebras. Benjamin, New York (1969)
85. Szczesny, M.: Colored trees and noncommutative symmetric functions. Electron. J. Combin. **17**, Note 19, 10 pp (2010)
86. Thomas, G.: On Schensted's construction and the multiplication of Schur functions. Adv. Math. **30**, 8–32 (1978)
87. Zhao, W.: A family of invariants of rooted forests. J. Pure Appl. Algebra **186**, 311–327 (2004)

Index

K. Luoto et al., *An Introduction to Quasisymmetric Schur Functions*,
SpringerBriefs in Mathematics, DOI 10.1007/978-1-4614-7300-8,

© Kurt Luoto, Stefan Mykytiuk, Stephanie van Willigenburg 2013